Guidebook for the Scientific Traveler

THE SCIENTIFIC TRAVELER

Duane S. Nickell, Series Editor

The Scientific Traveler series celebrates science and technology in America by highlighting places to visit of interest to educators, vacationers, and enthusiasts alike. Each book gives readers an introduction to the stories behind the sites, museums, and attractions related to topics like astronomy and space exploration, industry and innovation, geology and natural science. Doubling as a guidebook, each provides readers with useful and practical information for planning their own science-themed trips across America.

Guidebook for the Scientific Traveler

VISITING ASTRONOMY AND SPACE EXPLORATION SITES ACROSS AMERICA

DUANE S. NICKELL

Rutgers University Press
New Brunswick, New Jersey, and London

Library of Congress Cataloging-in-Publication Data

Nickell, Duane S.
  Guidebook for the scientific traveler : visiting astronomy and space exploration sites
    across America / Duane S. Nickell.
    p. cm. — (Scientific traveler)
  Includes index.
  ISBN 978-0-8135-4374-1 (pbk : alk. paper)
  1. Astronomy—Popular works. 2. Astronomy—Miscellanea. I. Title.
  QB44.3.N53 2008
  520.973—dc22

                                                                          2008000881

A British Cataloging-in-Publication record for this book is available from the British Library.

Visit our Web site: http://rutgerspress.rutgers.edu

Manufactured in the United States of America

This book is dedicated to my parents,
Anna June and Carl Duane "Red" Nickell.

CONTENTS

PREFACE

We are privileged to live in a modern world that is profoundly shaped by science and technology. Our everyday lives depend heavily on conveniences derived from advances in science and the technological applications that follow them. Our information, transportation, communications, and health care systems are deeply rooted in science. Even our democratic system of government has its origins in The Enlightenment when reason and rationality, as demonstrated by the science of Galileo and Newton, were applied to government. Most significant, we owe our understanding of how the universe works and of our place in it to the scientific enterprise. Science stands at the pinnacle of human intellectual achievement, and American science is the envy of the world. Our citizens have won far more Nobel Prizes in the sciences than any other country, and the best students from around the world flock to our universities for their scientific training. Yet the vast majority of the population has no formal scientific training beyond a few required science courses in high school and college. Relatively few people have careers in science and technology. But, of course, just because your job doesn't directly involve science doesn't mean you can't enjoy it. Just as music is not just for musicians, science is not just for scientists. Popular interest in science is high and should be encouraged. To experience and appreciate music, people go to concert halls, opera houses, and other music venues. Where are the places everyday people—nonscientists—can go to experience and appreciate science?

This is the first in a series of books that answers that question. This book series celebrates science and technology in America by listing, describing, and providing background information on scientific and technological points of interest across the United States, places that conventional travel guides tend to ignore. Some of these sites are obvious and others obscure. In addition to choosing the sites based on their scientific or historical significance, the central practical criterion for selecting a site is its accessibility to the public. Some sites, especially those on military and government property, may require that visitors reserve a spot for a tour well in advance. A few very notable sites are not publicly accessible, but they can be made accessible with minimal effort. I hope these sites eventually open their doors to the public.

The first book in this series focuses on scientific sites related to astronomy and space exploration. There are simply too many sites to list, much less describe, so I have been very selective and included only the most significant sites in each category. I have also made an effort to search out sites that are not well known. I have personally visited about half of the sites, but time and resources have precluded me from visiting all of them. The reader may quibble with the inclusion or exclusion of some sites. In particular, scientific purists may object to inclusion of Roswell and Area 51 in a book of purportedly scientific sites. Fear not! The chapter on these sites is written from a highly skeptical point of view. Like it or not (and I assure you I don't), these sites have captured the public imagination, and many people associate space travel and exploration with UFOs. I hope that my approach to these sites counters the uncritical view that one finds in the mass media, especially television. Besides, many scientific travelers may have an interest in pseudoscientific sites as well.

I would strongly encourage readers to contact me if they know of sites that may be worthy of inclusion in possible future editions of this book. I have striven to be factually accurate and correct. I regret any errors that may appear in this book and take full responsibility for them. Information regarding hours and admissions fees was accurate at the time of writing, but, of course, they may have changed since 2008. Please check the individual websites for up-to-date information. I have only included driving instructions for a few of the more isolated locations. Again, most websites have maps and directions as well as information regarding disabled access. Feel free to send any comments, corrections, critiques, and site suggestions directly to me at duane_nickell@yahoo.com.

In addition to listing the sites, the book provides background information on each location so that these special places may be more deeply appreciated. Every effort has been made to present the background information in a way that is comprehensible to the lay person. If the site is a museum, then I describe some of the "must see" exhibits so you know what to look for and won't miss anything. I recall finding the famous Martian meteorite ALH84001—the one that caused quite a stir a few years ago when NASA scientists claimed the rock contained microfossils of Martian bacteria—in the Smithsonian's Natural History Museum. Yet because of its inconspicuous location and signage, people were walking right by. I try to alert you to the little gems hidden in the big museums.

Understanding the scientific and historical significance of a place makes a visit a much more meaningful and enjoyable experience. For example, some may visit the Meteor Crater in Arizona and walk away disappointed, muttering to themselves, "It's just a big hole in the ground!" Ah, but if you know that big hole was made by a giant rock from outer space slamming into the Earth at nearly 30,000 mph, releasing the energy of 150 atomic bombs, and if you know that a much bigger rock vanquished the dinosaurs millions of years ago and that another giant rock could some day wipe out the humans, well . . . doesn't that shine a new light on that big hole?

I hope that this book may provide the reader and the traveler with a deeper appreciation for science, scientists, and the beauty of the natural world as revealed by science. Although this book is written for an adult audience, I hope that parents will take their children to some of these places. Who knows, maybe the child's curiosity will be piqued by the visit, and maybe he or she will consider a scientific career. Our country certainly needs to produce a bigger crop of scientists and engineers; our economy depends on it. My own interest in science was stirred, at least in part, by looking through a telescope at a local university.

Finally, I hope the reader has the opportunities to visit a few of the places described in this book. You need no more justification than simply the joy of being there—the pleasure of standing in a place where scientific history was made or seeing a telescope or a spacecraft that changed history. I remember standing on Mercer Street in Princeton, New Jersey, and looking at the house where Einstein lived—that was nothing short of a religious experience for me! But if you can't make it to these places, I have tried to include enough background scientific and historical material to keep even the armchair traveler entertained. Real or vicarious, let the journey begin!

ACKNOWLEDGMENTS

I owe a huge debt of gratitude to the following people, without whom this book would not have been possible. Thanks to my brother Christopher Shea Nickell, my aunt Lucy Metcalf, and my dear friend Pamela Nickell for their constant support and encouragement through the years. Thanks to my beautiful daughters Anna and Sarah for putting up with their dad being on the computer all the time. Thanks to my editor, Doreen Valentine, for her enthusiasm for this project and her patience in answering numerous questions from a first-time author. Thanks to Patsy Metcalf for her hospitality while I was visiting some of the sites. Special thanks to Karen Markman for serving as my sounding board throughout this project; she offered valuable input and suggestions regarding the manuscript. I would also pay tribute to the late Carl Sagan and Isaac Asimov, whose books kept me company on many lazy summer afternoons during my youth. Most of all, I would like to thank my parents Anna June and Carl Duane Nickell who instilled in me a love of learning that will never be extinguished.

Guidebook for the
Scientific Traveler

1
Native American Astronomy

For the earth he drew a straight line,
For the sky a bow above it;
White the space between for day-time,
Filled with little stars for night-time;
On the left a point for sunrise
On the right a point for sunset
On the top a point for noontide,
And for rain and cloudy weather
Waving lines descending from it.

From *Hiawatha* by
Henry Wadsworth Longfellow

n today's world, most of us rarely give the night sky more than an occasional glance. We live in cities and towns where the glow of electric lights overwhelms the twinkling of the stars. In the evenings, we dawdle away our time inside our snug and comfortable homes amusing ourselves by watching television or surfing the web. But there was a time before electricity and television and computers—a time when the distractions of modern life were absent. A thousand years ago, the best show in town was the night sky, and our ancient ancestors were regular viewers.

The two most prominent objects in the sky are the Sun in the daytime and the Moon at night. The dependable Sun provides light and warmth, while the Moon coyly performs its monthly cycle of waxing and waning. As they watched, our predecessors began to notice patterns in the motions of the Sun and the Moon. Every day, the Sun emerges from beneath the eastern horizon in the morning, arcs across the sky, and disappears below the western horizon at dusk. But the exact point along the horizon where it rises and sets shifts with the seasons. During the cold winter months, the Sun rises and sets farther south along the horizon, and, as summer approaches, the rising and setting Sun moves farther north. The day when the Sun reaches its farthest point south on the horizon is called the winter solstice; this first day of winter, December 21, has the least amount of daylight. The day when the rising and setting Sun reaches its northernmost point on the horizon is called the summer solstice, June 21; this first day of summer has the greatest amount of daylight. On two days each year the Sun rises and sets at a point exactly halfway between the northernmost and southernmost extremes. These days, equally divided into twelve hours of daylight and twelve hours of dark, are called the equinoxes. The vernal equinox marks the Sun moving along the horizon from south to north on March 21, the first day of spring. The autumnal equinox marks the Sun moving from north to south on September 21, the first day of autumn.

This back-and-forth motion of the Sun along the horizon is similar to the motion of a pendulum. Just as a pendulum moves fastest through the bottom of the swing and comes to a momentary stop at the top, the Sun changes its position along the horizon most swiftly at the equinoxes and comes to a momentary stop at the solstices. In fact, the word *solstice* means "stand still" in Latin. This oscillating motion of the Sun gave ancient Americans a way of devising a simple but sufficiently accurate horizon calendar. By taking advantage of natural horizon markers such as mountain peaks or rock formations, or by constructing artificial markers such as windows or poles, they could mark the points on the horizon that corresponded to the solstices and the equinoxes. Other markers might indicate the appropriate times for planting and harvesting crops or for important festivals and ceremonies. Of course, we now understand that the back-and-forth dance of the rising and setting Sun along the horizon happens because the earth's axis is tilted away from the Sun during winter in the northern hemisphere, so the Sun appears farther south in the sky. During summer in the northern hemisphere, the earth is on the opposite side of its orbit around the Sun, and now the axis is tilted toward the Sun so the Sun appears farther north.

Ancient Native American observers also watched the Moon complete its cycle from no Moon to a crescent of light, to a half circle, to a full circle, and back to no Moon, in a period of about 28 days; they noted the fact that the

Moon completes about twelve of these cycles each year. Many tribes gave to these cycles names, like "laying geese" or "coming caribou," a reflection of natural events happening at the time. Every month the rising Moon swings between a northern and southern extreme on the horizon, just as the Sun does every year. But owing to small cyclical variations in the Moon's orbit, the exact points of these extremes changes from month to month. When the Moon reaches its absolute maximum northern or southern position along the horizon, it has a "standstill" similar to the Sun at the solstice. The period of this lunar standstill cycle is 18.61 years. As described in the entries that follow, there is some evidence that the Ancestral Puebloans (formerly known as the Anasazi) of the American southwest knew of this lunar stand-still cycle as did prehistoric inhabitants of northern Europe.

In addition to the Sun and the Moon, Native Americans kept a close watch on the stars. The Navajo painted hundreds of stars on the roofs of caves. Today, about a hundred such "Star Ceilings" have been identified. The Pawnee people, especially a subtribe known as the Skidi, drew star charts, built shrines dedicated to certain stars, and used the appearances of stars as a calendar. They attached particular importance to Mars and carefully tracked its wanderings across the sky. The roofs of Pawnee earth lodges were supported by four poles that symbolized the four stars that held up the sky. The identity of these four stars is unknown, but each was associated with a particular color, animal, season, and type of corn.

The early Native Americans showed no great interest in carving the sky into constellations, but a few of the most prominent star groups did have legends associated with them. Many of these stories share the common theme of pursuer and pursued. This makes sense in light of the fact that hunting was a matter of survival, and the chase seemed represented in the eternal wheeling of the stars across the arc of the sky. For example, sub-groups of the Iroquois, Inuit, and Yupik envisioned the Big Dipper as a bear, represented by the four stars that make up the cup of the dipper, being hunted by the three stars in the handle. The Blackfoot tribe, reversing the story, claimed that the seven stars of the Dipper were seven youths who had been chased into the sky by a bear. The constellation we call Orion repre-sented a hunter to a number of Native American tribes just as it did to the ancient Greeks. Some southwestern tribes gave the hunter the burden of car-rying the Sun on his back. In an unusual interpretation, the Pima people saw Orion as a coyote in hot pursuit of the Pleiades, seven women who had com-mitted the crime of cannibalism. To the Blackfeet, the semicircle of stars known as Corona Borealis represented the abdomen of a spider and the brightest stars in the neighboring constellation of Hercules as filaments of the web.

Although the early Native American population was divided among dozens of tribes, each with its own unique set of myths and legends about the sky and the heavens, a few ideas were common to many of them. One shared concept was the way that early Native Americans thought of time itself. There was no need to know the exact time of day or the exact day of the year. Time was told by natural events: the behavior of animals, the changing colors of leaves, the position of the Sun, and the cycles of the Moon.

The Native Americans believed that the four cardinal directions—north, south, east, and west—were sacred. Consequently, many tribes oriented their structures and laid out their communities along these directions. People from various cultures around the world have designed their homes and temples to reflect the architecture of the cosmos. The domes of the Indian *stupa,* the Tibetan *chorten,* and even St. Peter's Basilica mimic the celestial sphere. Native Americans were no different. The Ancestral Puebloans, for example, performed their religious rituals and ceremonies in circular subterranean chambers called *kivas.* Most kivas were less than 25 feet in diameter (the so-called "great kivas" were much larger), dug deep enough to allow the occupants to stand upright. The kivas were covered by a roof with a hole cut into it to serve as an entrance. A ladder provided access to the floor. The architecture of the kiva is symbolic of the creation story and cosmology of the Ancestral Puebloans. According to modern Pueblo and Hopi tradition, the first humans lived deep in the underworld. Humans migrated from one underground realm to another until they emerged into the fourth and final world, the surface of the Earth. The roof entrance and ladder symbolized this emergence, and the kiva itself represents the *sipapu,* the place where humans first emerged from the underworld. Most kivas contain a symbolic sipapu in the floor located just north of the center of the kiva. This sipapu represents the emergence of the ancestors from the second into the third world. The four posts supporting the roof of the great kivas may represent the trees planted in the underworld for the people to climb up on. The walls and roof represent the circle of the sky and the roof beams represent the Milky Way.

The sites described below are places where you can go to gain a deeper appreciation for and understanding of the astronomical knowledge and practices of prehistoric Native Americans. They are grouped together according to tribe starting with an unknown tribe from the Late Woodland culture, then on to the Plains Indians, and ending with several sites built by the Ancestral Puebloans.

Cahokia Mounds, Illinois

This site across the Mississippi River from St. Louis was first inhabited around 700 A.D. by Native Americans of the Late Woodland culture who chose the site for its fertile, easy-to-till soil. By 1000 A.D., Cahokia was the center of trade for the Mississippian culture. At its peak, the population was 15,000, making it the largest city ever built north of Mexico before Columbus. After 1200 A.D., the population began to decrease, and the site was completely abandoned by 1400. The Cahokia Indians, a subtribe of the Illini, settled here in the late 1600s, and the site is named after them rather than the inhabitants of a thousand years ago, whose language and ethnicity remains a mystery; we don't even know what they called themselves.

Today, the ancient city is best known for its 80 surviving earthen mounds, many of which took the shape of truncated pyramids with a square base. Civic leaders built their homes atop these pedestals as a symbolic reminder of their elevated societal status. The most massive of the mounds is the 100-foot-tall Monks Mound, so-named because French monks stumbled upon the mounds in the mid-1700s and named the biggest one after themselves. With a base that covers an area of fourteen acres, Monks Mound is the largest prehistoric earthen structure in the Americas. The original network of 120 mounds and community plazas at Cahokia required that an estimated 55 million cubic feet of earth be hauled in woven baskets using human muscle power. The mounds stand as a testimony to the engineering and construction skill of these unknown ancient people.

The Cahokian site of most interest to the astronomically inclined traveler is affectionately known as Woodhenge, after its more famous stone cousin in England. Woodhenge was discovered in 1961 by archeologist Dr. Warren Wittry. At the end of a summer of intense excavation in an effort to harvest archaeological information from land that was destined for use as an interstate highway (the highway was later rerouted), Wittry was studying maps of the excavated site and noticed that large bathtub-shaped, sloping pits were arranged in several circular arcs. He realized that the pits were

An artist's rendering of the construction of Woodhenge.

probably used to slide large wooden posts into holes so that a circle of timbers 20 feet tall could be erected. Further excavation unearthed more pits and postholes until the evidence suggested that as many as five separate Woodhenges were constructed between 900 and 1000 A.D. Bits of red cedar, a wood sacred to the Cahokians, were found in some pits along with red ocher pigment, which indicates that the posts may have been painted.

As to the purpose of the wooden circles, Wittry hypothesized that they may have been used as a sort of solar calendar. He discovered that from an observation point near the center of Woodhenge 3 (also called Circle 2), two of the posts line up with the point on the eastern horizon where the Sun rises on the mornings of the summer and winter solstices. A third post halfway between these two posts marks the point on the horizon of sunrise on the equinoxes. This equinox pole also lines up with the front of Monks Mound so that at sunrise on the equinoxes, the Sun emerges dramatically from behind the mound. A pit has been discovered near the winter solstice post. Some researchers have suggested that the pit may have held ceremonial fires lit to warm the Sun and persuade it to reverse its direction along the horizon so that another yearly cycle could commence. Perhaps the winter solstice was the Cahokian equivalent of New Year's Day. You may be bothered by the fact that all these alignments are based on an observation point that is not at the exact center of the circle. But interestingly enough, a posthole was found five feet east of the center of the circle—the precise spot required for the alignments. A pole may have been erected here that

Cahokia Mounds State Historic Site

The reconstructed Woodhenge. The central observing post can be seen in the foreground with the circle of posts in the background.

supported an elevated observation platform; a sun priest may have stood there to reverently track the motion of the Sun.

Woodhenge 3 has 48 posts and only a few served as solar horizon markers. What were the others used for? Archeologists speculate that some posts may have marked important ceremonial or festival dates perhaps related to the planting and harvesting of crops. Other posts may be aligned with the Moon or bright stars. Another possibility is that the remaining posts were raised simply to complete the circle in which sacred ceremonies were held. At least one archeologist has suggested that in addition to its use as a solar calendar, Woodhenge served as an engineering "aligner" to aid in determining the proper placement of the earthen mounds.

Woodhenge, reconstructed in 1985 at the original location of one circle, is the one site at Cahokia that scientifically minded travelers do not want to miss. Special sunrise lectures and observances are held at Woodhenge near the equinoxes and solstices. Check the website for dates and times.

Visiting Information

Cahokia Mounds State Historic Site is located eight miles from downtown St. Louis and is easily accessible from Interstates 55/70 and 255. Just follow the signs. The site is open daily from mid-April through Labor Day and is open Wednesday through Sunday the rest of the year. The site is closed on most holidays. The grounds are open from 8:00 A.M. to dusk. Admission to the site is free, although a donation of $2 for adults and $1 for children is requested.

Your first stop should be the excellent Cahokia Mounds Interpretive Center where you can pick up some maps and brochures, plan your visit, browse through the gift shop, grab a snack, and learn about Cahokian culture, politics, economics, and religion. The "City of the Sun," a 15-minute video introduction to the history of Cahokia, is shown every hour in the auditorium. The center also features a re-creation of a Cahokian village and an exhibit on Woodhenge and the sun calendar. Guided one-hour tours are offered during the summer months at 10:30 A.M. and 1:30 P.M. on Wednesdays through Saturdays and at 12:30 P.M. and 2:30 P.M. on Sundays. The morning and afternoon tours go to different areas of the site. During April, May, September, and October, the tours are given on weekends at 1:30 P.M. There is also a self-guided

> Website: www.cahokiamounds.com
> Telephone: 618–346–5160

audiocassette tour available and a walking tour guidebook. Woodhenge is located along the state highway that divides the park. Parking is available at the Woodhenge site.

Bighorn Medicine Wheel, Wyoming

Built by the Plains Indians, medicine wheels are patterns constructed from small rocks. The usual pattern consists of a circle of rocks centered on a large pile of similar rocks called a *cairn*. Most medicine wheels feature lines of stones radiating outward from the central cairn to the outer circle like the spokes of a wheel. The number of spokes is variable and they are not evenly spaced. Sometimes the spokes extend beyond the outer ring and sometimes, they originate from the outer ring. Often, medicine wheels feature additional stone cairns outside of the main wheel. Sometimes, these ancillary cairns are attached to spokes or to the main wheel itself and sometimes they are completely separate from the main structure. More than 70 medicine wheels have been identified in North Dakota, Montana, and Wyoming in the north central United States and in Alberta and Saskatchewan in south central Canada; the highest concentration is in Alberta. The wheels are usually found on high ground where there is a clear, unblocked, 360° view of the horizon. The estimated age of these medicine wheels ranges from a few

The Big Horn Medicine Wheel in Wyoming.

hundred to a few thousand years; one of the oldest dates to more than 4,500 years ago. Use of the descriptive term "medicine" refers to the supernatural and the mysterious, similar to the usage in "medicine man." Although archeologists aren't exactly sure what function the medicine wheels served, they were probably used for various ceremonial, ritualistic, spiritual, and, perhaps, astronomical purposes.

The prototypical medicine wheel is the Bighorn Medicine Wheel located near the summit of Medicine Mountain, one of the highest peaks of the Bighorn range at an elevation of nearly 10,000 feet. At first glance, the wheel may disappoint visitors because it is a rather crude structure made from whitish, flat rocks. The central cairn, about 2 feet tall and 12 feet wide, is hollowed out in the center. From the central hub 28 spokes radiate outward, terminating at the outer circle with a radius of about 35 feet. The outer ring is not perfectly circular but more of an imperfect ellipse. At irregular intervals six additional U-shaped cairns decorate the circumference of the wheel; four cairns lie outside the ring, and one is situated so that the bottom of its U touches the inside of the ring, and another lies entirely outside the ring at the end of an extended spoke. The U-shape of the cairns points in various directions.

Archeologists estimate that the Bighorn Medicine Wheel is between 200 and 400 years old. They also note that the design of the medicine wheel very closely resembles the plan of a Cheyenne medicine lodge. These lodges were used to celebrate the sun dance ceremony, the most important and frequent summer ritual practiced by the Plains Indians. The round lodges had a central post, which might be symbolized by the central cairn of the medicine wheel. Moreover, the lodges had 28 rafters that radiated outward from the central post in the same way that the wheel has 28 stone spokes. Finally, some lodges featured an altar on the western side and the Bighorn wheel has a cairn near its western edge. These similarities suggest that the medicine wheel was a two-dimensional representation of a medicine or sun dance lodge built out of stone because it was above the timberline where wood was not readily available. This lends credence to the belief that the purpose of the wheels was largely ceremonial.

In the 1970s, the solar astronomer Dr. John A. Eddy visited the Bighorn Medicine Wheel and discovered that lines of sight extending through various pairs of the seven cairns coincided with certain astronomical events. Most significantly, if one stands at the distinctive cairn that lies outside of the main ring and sights through the center of the central cairn, you are looking at the point on the horizon where the Sun rises on the summer solstice. Standing at another peripheral cairn and again sighting through the central cairn, you look to the point on the horizon where the Sun sets on the summer solstice. Dr. Eddy then found astronomical alignments using all but one of the other cairns. First, an alignment through two of the outer cairns points to the spot on the horizon where the bright star Aldebaran would have risen just before dawn within a couple of days of the summer solstice between 1600 A.D. and 1800 A.D., a time period in agreement with the estimated age of the wheel. (The exact positions of the stars shifts over time owing to changes in the direction that the Earth's axis points.) A line drawn through two other cairns is directed to the point where the star Rigel would have risen shortly before dawn 28 days after the summer solstice during the same time period. A final alignment through a pair of cairns coincides with the point on the horizon where the star Sirius would have risen 28 days after Rigel. All three of these prominent stars (Sirius is the brightest star in the sky) have some significance in the lore of the Plains Indians.

Imagine an observer at the medicine wheel several hundred years ago. He would see Aldebaran rise to announce the approach of the summer solstice, then a month after that, the rise of Rigel, and a month after that, the rise of Sirius toward the end of August. The appearance of Sirius near the end of summer perhaps warned the observer that it was time to come down from the mountain.

Eddy was annoyed at not finding an alignment that used the final cairn. However, another astronomer, Jack Robinson, found an alignment using the seventh cairn that marked the dawn rising of the bright star Fomalhaut approximately one month before the solstice. Of course, it is possible that all these astronomical alignments are coincidental. However, Eddy later investigated the Moose Mountain Medicine Wheel in Saskatchewan and discovered the same astronomical alignments. This increases the likelihood that the alignments were intentional.

Visiting Information

The Bighorn Medicine Wheel has been designated as a National Historic Landmark. It is located along Wyoming State Highway 14A between Burgess Junction and the western edge of the Bighorn National Forest, about 100 miles east of Yellowstone National Park. The highway snakes up the steep western slopes of the Bighorn Mountains in a series of switchbacks. The site is open to the public from 8 A.M. until 5 P.M. from June 15 through September 30. It is closed the rest of the year due to snow. There is no admission fee. The site remains ceremonially important to more than 60 tribes and is occasionally closed for a short periods of time for Native American ceremonies. A small visitor's center and parking lot are located at the entrance to the site. Here, you can pick up some informative literature and begin your 1.5-mile hike up to

Websites: wyoshpo.state.my.us/medwheel.htm
http://wyomingtourism.org
Telephone: 307–548–6541

the wheel. A rope fence attached to wooden posts encircles the wheel. Visitors are asked to walk to the left around the wheel out of respect for the tribes that still use the wheel. (Walking around to the left mirrors the motion of the Sun across the sky.)

Medicine Wheel Park, Valley City, North Dakota

The construction of this unique park began in 1992 as a classroom project led by Dr. Joe Stickler, a science professor at Valley State University. The project quickly expanded to involve the local community. Today, the 30-acre park includes replicas of two types of solar calendars used by prehistoric Native Americans, a scale model of the solar system, an Indian Burial Mound, a perennial flower garden, and nature trails.

The park's main attraction is the Medicine Wheel, a rock sculpture modeled after Wyoming's Big Horn Medicine Wheel. Although the dimensions of this young Medicine Wheel are nearly the same as those of its older inspiration, it has a much sharper and clearly defined appearance. Like the Big Horn Medicine Wheel, this replica has 28 spokes radiating from its center representing the 28 days of the lunar cycle. Six of the spokes extend beyond the wheel and are aligned with the points on the horizon where the Sun rises and sets on the equinoxes and solstices.

Near the Medicine Wheel, another type of ancient solar calendar consists of a 38-foot-tall utility pole and three boulders. Here, the seasons are tracked by the length of the shadow cast by the pole. The pole cast longer shadows in the winter when the Sun executes a lower arc across the sky and shorter shadows in the summer when the Sun executes a higher arc. The shadow is at its absolute longest on the winter solstice when the Sun's path is lowest with respect to the horizon. On this day, the top of the shadow reaches the farthest boulder, 100 feet from the base of the pole. The shadow will attain its absolute minimum length on the summer solstice when the Sun's path is highest with respect to the horizon. On this day, the top of the shadow reaches the nearest boulder, only 15 feet from the pole. On the two equinoxes, the shadow falls on the middle boulder, 38 feet from the pole. Using this type of "meridian calendar," the thirteenth-century Chinese astronomer Guo Shoujing was able to measure the exact length of a year to the nearest minute.

The park also features a scale model of the solar system, which originates at the center of the Medicine Wheel and is decorated with reddish gravel to represent the Sun. In this model, one foot is equivalent to approximately three million miles, and the positions of the planets are marked by boulders. The earth's orbit coincides with the main circle of the wheel, while the orbits of Mercury and Venus fall inside the circle, and Mars's orbit lies just beyond the circle. The boulder representing Jupiter is found near the parking lot. The path leading east from Jupiter takes you past Saturn, Uranus, Neptune, and, finally, the dwarf planet Pluto, a third of a mile from the center of the wheel. On this scale, Proxima Centauri, the nearest star aside from the Sun, would be located in LaPaz, Mexico.

Visiting Information

Medicine Wheel Park is located on the campus of Valley City State University in Valley City, North Dakota, about an hour's drive west of Fargo. From I-94, take exit 292, drive north 0.2 miles, and turn right (east) onto Winter Show Road. After about a half mile, you'll see the sign for Medicine Wheel Park. The park is open year round 24 hours a day and there is no admission fee.

Website: http://medicinewheel.vcsu.edu

Chaco Canyon National Historical Park, New Mexico

A thousand years ago, this isolated desert valley in what is now northwestern New Mexico was home to a thriving settlement of several thousand Ancestral Puebloans. Today, visitors can come here to see some of the most impressive Native American ruins in North America. The inhabitants lived in some 400 settlements, including larger towns and smaller villages, scattered across the canyon and throughout the neighboring region. The larger towns were built around a large pueblo with oversized rooms surrounded by smaller villages. The Chacoans were skilled builders and employed several basic styles of masonry. The earlier buildings were constructed with simple walls one stone thick held together by generous quantities of mud mortar. The later, multistory buildings feature tapered walls with thin stone veneers filled in with thick cores of rubble.

Archeologists are amazed by an extensive system of more than 400 miles of roads connecting 75 of the communities. The roads, laid out along straight lines, average 30 feet in width. Not merely foot trails, these roads, made visible by prolonged wear, show signs of careful engineering. The road beds have been excavated below the surface and are bounded by earthen berms. The roads identify Chaco Canyon as both a ceremonial center and the hub of a vast regional network.

Chaco Canyon enjoyed 150 years of success until an extended drought hit the San Juan Basin between A.D. 1130 and 1180. There is some controversy concerning the causes of the decline of Chaco Canyon, although most agree the drought was certainly a factor. Some scientists believe that outsiders may have introduced new rituals to the region that caused a schism between the tribes. Others point to overextended resources and an overtaxed environment that forced some to seek greener pastures. Nevertheless, while they were there, the Chacoans kept a careful watch on the sky. Six sites at Chaco Canyon have been identified by archeoastronomers that possibly reveal Chacoan interest in the Sun and the sky: Pueblo Bonito, the Fajada Butte "Sun-Dagger," Casa Rinconada, the supernova pictograph at Penasco Blanco, a sun-watching station at Wijiji, and a solar eclipse pethoglyph at Una Vida.

At the base of a 100-foot high mesa sits Pueblo Bonito (Beautiful Town), the largest prehistoric southwest Native American dwelling ever excavated. This gigantic pueblo is a D-shaped 600-room multistory apartment house

Ashok Rodrigues/iStock

Pueblo Bonito at Chaco Canyon.

covering an area of over three acres. Little remains of the upper floors, but several sections of five- to six-story-tall walls are still standing. The layout of the building is carefully oriented along the four cardinal directions. For starters, the large central plaza is split in two by a low wall that is aligned to within three-quarters of a degree of true north and south. The great kiva that connects to this center wall is also closely aligned with north and south. The western half of the outer wall that forms the straight leg of the "D" is almost perfectly aligned to true east and west.

Pueblo Bonito's walls feature a number of corner windows that are not usually included in Puebloan architecture. Sunrise on the summer solstice can be easily viewed from two of the windows that face east. From these windows, the motion of the Sun along the eastern horizon can be tracked until late October. A little later in the year, a thin pencil of light appears on the wall opposite the window at sunrise. The shaft of light widens and moves northward along the wall as the winter solstice approaches. On the solstice, a sharp square patch of light appears in the corner of the room. It is possible that Chacoan Sun priests marked out a crude solar calendar on the plaster that used to coat the walls of the room. However, some evidence suggests that outer walls may have blocked sunlight from entering the rooms.

Of additional interest is the fact that that one researcher has claimed that Pueblo Bonito's design successfully utilizes passive solar energy. For example, during the winter when the Sun is lower in the sky, the curved outer wall acts like a mirror and reflects light and heat into the central plaza. In the summer, the rooms are shaded by the walls of the canyon. In spite of wildly fluctuating temperatures in the desert outside, the temperature in the rooms is kept fairly constant by the thick walls and reflection of heat and light by the cliff walls.

Almost directly across from Pueblo Bonito lies the great kiva known as Casa Rinconada. With a diameter of 63.5 feet at floor level, Casa Rinconada is the largest kiva in Chaco Canyon and is among the largest ever built by the Ancestral Puebloans. Around the interior wall are 28 regularly spaced niches along with 6 larger, irregularly spaced niches placed a bit lower along the wall. The 28 higher niches could represent the number of nights it takes for the Moon to return to the same position among the constellations. The kiva is aligned with the four cardinal directions with exquisite precision. The north and south entrances are aligned to within a third of a degree of true north. Lines drawn through all but one opposite pairs of the 28 wall niches run directly through the center of the kiva, and one pair establishes a near perfect east-west line. The holes that once supported the roof posts form a square in which each side is within a half a degree of the exact cardinal directions.

Researchers have discovered that, during a four- to five-day period centered on the summer solstice, sunlight entering a narrow window shortly after sunrise falls squarely into one of the six irregularly spaced compartments. Whether this is an intentional or accidental alignment remains unclear. One problem is that a room may have existed directly outside the critical window. This room would, of course, have blocked the sunlight from the window. However, the room may not have been present throughout the useful life of the kiva. It is also possible that one of the roof support beams could have prevented sunlight from reaching the niches. Finally, some evidence suggests that a screen, once in front of the niches, would have blocked the sunlight.

Another possible astronomical marker in Chaco Canyon can be found just outside the great house called Penasco Blanco. At the base of the mesa occupied by these dwellings, on the underside of a small overhang, is a pictograph containing these symbols: a large star, a crescent Moon, and a handprint. Some archeoastronomers believe this pictograph depicts the

Robert Fullerton/Shutterstock

The Supernova petroglyph at Chaco Canyon. The crescent moon is at the right and the handprint is at the top.

supernova of 1054. A supernova is a massive star that explodes. This super-nova happened relatively nearby, only about 4,000 light-years from Earth. Astronomers estimate that this supernova was five or six times brighter than the planet Venus, which, aside from the Sun and the Moon, is the brightest object in the sky. Although its brilliance gradually diminished with time, this supernova was visible during the day for 23 days and could be seen in the night sky with the naked eye for three months. This rare and spectacu-lar astronomical event could not have been missed by even casual observers of the sky. The Chinese, Japanese, and Arab civilizations took note of the event, although curiously, there is no record of it in Europe. Today, the rem-nant of this supernova is called the Crab Nebula, a vast cloud of gas and dust in the constellation of Taurus the Bull.

What evidence is there that these symbols actually represent the 1054 supernova? Astronomers can calculate the precise phase of the Moon and exactly where the supernova would have been located in relation to it. The pictograph correctly shows the relative position of the supernova and the crescent Moon on the morning of July 4, 1054, when the supernova first appeared. The hand, perhaps, is the artist's signature. In fact, every 18.5 years, the Earth and the Moon return to approximately the same positions they occupied in July of 1054. If, on one of those dates, you point a telescope to the position in the sky indicated by the star symbol in the pictograph, you will see the Crab Nebula. Unfortunately, the age of the pictograph cannot be determined. The best archeologists can do is to infer the age of the pictograph from the time during which the Chaco Canyon community was flourishing. Tree ring data along with other archaeological evidence all point to the middle of the eleventh century as the golden age of the Chaco Canyon civilization.

Beneath the supernova pictograph, on a vertical rock wall, are three concentric circles about a foot in diameter. To the right of the circles flames appear to be painted in red paint that has faded with time. In fact, the paint is so faint that the flames don't usually show up in black-and-white photographs. Interestingly enough, Halley's comet made an appearance in 1066, just a dozen years after the supernova of 1054. Some researchers have suggested that the circles and flames might represent the comet. Perhaps both of these unusual astronomical occurrences were purposely recorded in the same place. Again, the evidence is inconclusive and some experts claim the circles are merely a symbol of the Sun.

At the extreme eastern end of Chaco Canyon are the ruins of Wijiji, one of the last towns to be built in the canyon. A short distance from the ruins, a staircase takes you to a ledge that runs along the rim of the canyon. This ledge is decorated with Ancestral Puebloan and Navajo paintings and carvings, including a faded Sun symbol. Further along the ledge, you come to a group of boulders, one of which is marked with crosses and spirals cut into the rock by the Chacoans. A few yards beyond this spot, look to the southeast across to the other side of the bend in the canyon rim. There, you see a natural rock pillar. If you stand here on the winter solstice, you see the Sun rise from directly behind the rock pillar. At sunset on the same day, the Sun lowers itself down into a natural cleft in the rock a short distance away. This arrangement has led some researchers to conclude that this spot may have been used by Chacoan sun-watchers.

One of the most famous Native American astronomical markings in the entire United States is at Fajada Butte here in Chaco Canyon. Fajada Butte, one of the most visible landmarks in the park, stands in magnificent isolation at a fork in the canyon near its eastern end and is impossible to miss. On a June day in 1977, an artist was sketching two spiral petroglyphs near the top of the butte for a survey of rock art in the canyon. The spirals were in a darkened area hidden behind three vertical slabs of sandstone that shade it from direct sunlight. Around noon, she noticed that a sliver of sunlight cut through the spiral. Upon further investigation, it was discovered that just before noon on days near the summer solstice, a dagger of light cuts through the center of the larger spiral. Near noon on the winter solstice, two daggers appear on either side of the larger spiral. At noon on the equinoxes, a shaft of light bisects the smaller spiral while a larger light dagger passes to the right of the center of the greater spiral.

This so-called Sun Dagger petroglyph also appears to mark significant lunar events. It turns out that the Moon's shadow bisects the spiral at moonrise during the minor northern standstill and just barely touches the left edge of the spiral during the major northern standstill. At both of these positions, a straight groove has been cut in the rock at the same angle as the Moon's shadow, which is different from the angle of the daggers of sunlight. Moreover, the larger spiral contains 19 grooves. It just so happens that the period of time between the major standstills is 18.61 years, and the period of time it takes for exactly the same phase of the Moon to occur on the same day of the year (the so-called "Metonic" cycle of the Moon) is 19 years. It is therefore possible that the Chacoans were aware of these lunar cycles as well as cycles of the Sun.

Unfortunately, the Sun Dagger no longer performs its light dance. In 1989, the slabs of rock shifted, and the effect was ruined. Today, special permission from the Park Service is required to climb the butte.

The final site of astronomical interest at Chaco Canyon is a large boulder near the Una Vida ruins. The boulder itself may have been a sun-watching station. If you stand on the northeast side of the boulder at a spot indicated by a spiral petroglyph, you see on the northeast horizon a rock in the shape of a sharp pyramid. At sunrise on and around the summer solstice, the Sun appears near this rock. On the south side of the boulder is something even more interesting. There, you will find a petroglyph of a solid disc with curving filaments emanating from it. This petroglyph can be interpreted as a depiction of a total solar eclipse with the disc representing the eclipsed Sun and the filaments representing structures in the solar corona called "helmet streamers," long, spiked cones apparent during an eclipse. To the upper left of the petroglyph is a solid dot that might represent the planet Venus, which often makes a spectacular appearance near the Sun during a total eclipse. But was there a total eclipse of the Sun visible from somewhere in the San Juan basin during the relevant time period? In fact, there were four: April 13, 840; July 11, 1097; June 13, 1257; and October 17, 1259. The date of the 1097 eclipse corresponds with the apex of the Chaco Canyon community. Although we can never be certain, it is plausible that a Chaco Canyon sun-watcher would have considered a rare total solar eclipse an event worthy of recording.

Visiting Information

Begin your visit at the Visitor Center where you can browse through exhibits and see films about Ancestral Puebloan culture. Here, you can pick up brochures for self-guided trails and inquire about ranger-guided walks and campfire programs that are available during the summer. The Visitor Center, including a bookstore, is the only place in the park where drinking water is available. The center is open daily from 8:00 A.M. to 5:00 P.M. and from 8:00 A.M. to 6:00 P.M. during the summer months. The trails are open from sunrise to sunset. A one-way road runs from the Visitor Center up one side of the canyon and down the other. Parking lots are located near the major pueblos. Of the sites described above, Pueblo Bonito and Casa Rinconada are easily accessible from the road. Backcountry hikes of several hours are required to reach Penasco Blanco and Wijiji.

This park is located in an extremely isolated area. There is no food, lodging, or gas available in the park. The nearest town is 60 miles away. During the week, supplies can be purchased at trading posts along Highway 44. There is a park campground with 64 sites about a mile from the Visitor Center. The camping fee is $10 per night and permits are available at the Visitor Center. The campground has tables, fire grates, and toilets.

The park is located in northwestern New Mexico. If you are coming from the north, turn off New Mexico Highway 44 at Nageezi and follow San Juan County Road 7800 for 11 miles to New Mexico 57. From there, the Visitor Center is 15 miles. From the south along I-40, take New Mexico 57 north for 44 miles at Thoreau. Two miles north of Crownpoint, NM 57 turns to the right. Continue east

> Website: www.nps.gov/chcu
> Park telephone: 505–786–7014
> 24-hour emergency telephone: 505–786–7060

on NM 57 until you get to a marked turn-off. A 20-mile stretch of dirt road leads you to the Visitor Center. Admission to the park is $8 per car. Be sure to call ahead at the number above and check the road conditions. The dirt roads can get muddy, especially after the frequent afternoon thunderstorms in the late summer.

Chimney Rock Archaeological Area, Colorado

About 100 miles north of Chaco Canyon, the magnificent natural twin pillars of Chimney Rock stand guard over the ruins of an Ancestral Puebloan community in which more than 200 homes and ceremonial buildings were occupied between 925 and 1125 A.D. The architectural style and masonry techniques link this settlement to Chaco Canyon. Perhaps the most mysterious and dramatic ruin is the Great House, which sits like a miniature Machu Pichu 1,200 feet above the valley floor and directly below the spires. The Great House holds two kivas and 35 ground-floor rooms, although a second-story floor may have added an additional 20 rooms.

But why build here? The location is difficult to reach and accessible only by crossing a long, narrow rock causeway. Water would have to be hauled up from the valley below, and no fertile soil could be used for growing crops. It appears that the Great House was built here simply to be close to the pillars; perhaps they served some spiritual or ceremonial purpose.

In 1988, a team of archeoastronomers led by Dr. McKim Malville of the University of Colorado discovered that for a period of about two years during each 18.6-year lunar standstill cycle, the Moon rises between the two rock spires, as seen by observers at the Great House. These scientists hypothesize that the building was constructed here at least in part because of the alignment of the rising moon and the pillars. Two significant episodes of construction support this theory: by analyzing logs found at the site the dates of construction coincide with the last two lunar standstills of the eleventh century, specifically in October 1075 and June 1094. Did the Ancestral Puebloans know about the 18.6-year standstill cycle? Maybe, and then again, maybe not. But they most certainly knew about the solstices. Every nineteen years (recall that this is the Metonic cycle of the Moon), the full Moon rises close to the time of the winter solstice. During the eleventh century, a full Moon occurred at or near the winter solstice in the years 1055, 1076, and 1095. These dates are within a year of the lunar standstill dates cited above. Thus, the most highly anticipated astronomical event viewed by the inhabitants of Chimney Rock may have been the rising of the full Moon between the pillars near the winter solstice.

But what about the Sun? If they used the pillars as reference points for the Moon, then surely the Ancestral Puebloans would have used them for the Sun as well. Malville has identified several sites that may have been used for sun-watching. First, about 2,000 feet southwest of and below the Great House is a partially dismantled stone circle containing a small stone basin carved into the rock. If you stand over the basin and site along the northern wall of the Great House, you see the Sun rise over the center of the wall on the summer solstice. From this same basin, the southernmost point on the Great House lines up with the point in the sky where the 1054 supernova would have appeared. Second, from a small tower located along the southeastern cliff edge below the spires the rising Sun could have been observed throughout the year. Mountain peaks and valleys along the eastern horizon could have served as horizon markers for a solar calendar. Finally, across and above the Piedra River to the west lies a long, high ridge with 12 sites. From the vantage point of each of these sites, observers can see the Sun rising through the pillars at various times of the year.

Visiting Information

Chimney Rock Archaeological Area occupies 4,100 acres of the San Juan National Forest and has been designated as a National Historic Site. It is located 17 miles west of Pagosa Springs, Colorado (or 42 miles east of Durango). To get there, take US Highway 160 to Colorado Highway 151. Take HW 151 south for 3 miles, and you see the entrance on your right. The Visitor Center is half a mile from the entrance. Chimney Rock is open daily from May 15 through September 30. It is therefore regrettable that visitors cannot see the Moon rise between the pillars at the winter solstice. The Visitor Center is open from 9:00 A.M. to 4:30 P.M. with guided tours leaving at 9:30 A.M., 10:30 A.M., 1:00 P.M., and 2:00 P.M. From June 15 to August 31, an additional tour is offered at noon. Tour prices are $8.00 for adults, $2 for children (ages 5 to 11), and children under five are free.

The tours start at the Visitor Center and last about 2.5 hours. The guide leads a car caravan to the upper parking lot at the top of the mesa where the approximately one-mile walking tour begins. The mesa is home to 16 individual ruins, 14 of which are residential. The tour starts with the Pit House, the Great Kiva, and some unexcavated sites along the Great Kiva Trail Loop. After a visit to the Ridge House, you walk across the narrow ledge to the restored ruins of the Great House Pueblo. Look back for a spectacular view of the mesa with the San Juan Mountains to the east and the Piedras River valley to the west. Look ahead to see the 300-foot-tall twin rock pillars of Companion Rock and Chimney Rock. An old forest service fire lookout house sits between the Great House and the pinnacles and looks rather out of place.

Of particular interest to the scientifically minded traveler are special astronomy-related programs offered at Chimney Rock. Monthly night-time full-Moon hikes take you to the highest part of the Great House Pueblo for a talk on the Ancestral Puebloans who lived here and

> Website: www.chimneyrockco.org
> Telephone: 970–883–5359 in season;
>   970–264–2287 out of season

their interest in astronomy. In addition, there is a summer solstice program on June 21 and an autumnal equinox program on September 21. Check the website for a program schedule.

Hovenweep National Monument, near Cortez, Colorado

Hovenweep (a Paiute/Ute word meaning "Deserted Valley") National Monument is home to six clusters of Native American ruins. The structures include kivas, D-shaped dwellings, and, most notably, square and circular towers. Most structures were built between 1200 and 1300 A.D. by Ancestral Puebloans, a farming culture that occupied the region from about 500 to 1300 A.D. Archeologists believe that during that time, the area had deeper soil, more moderate temperatures, and more rain than it does today. At its height in the late 1200s, the population of the Hovenweep area exceeded 2,500. An extended drought, possibly coupled with battles with hostile neighbors, overpopulation, and depletion of resources, forced the inhabitants to abandon the area around 1300 A.D. The first Europeans to stumble upon the ruins were Mormon colonists in 1854. In 1917–1918, a Smithsonian Institution ethnologist surveyed the ruins and recommended that they be protected. In 1923, President Warren G. Harding made Hovenweep part of the National Park System.

Hovenweep is best known for its unusual and spectacular towers, most of which are less than two stories tall and two meters in diameter. There is no archaeological consensus as to the function of the towers, but possible uses include storage facilities, defense against enemies, family dwellings, or civil buildings. Some towers are connected to a kiva via a tunnel, suggesting a ritualistic use.

At least one tower, Hovenweep Castle in the Square Tower Group, may have been used as a kind of solar calendar. In a corner on the western side

National Park Service Photo

Hovenweep Castle at Hovenweep National Monument.

of the castle, one room has three low entrances and two small window slots. Archeologist Ray Williamson discovered that at sunset on the summer solstice, sunlight enters through one window and forms a patch of light near the lintel stone of one low doorway. At the winter solstice, the Sun shines through the other hole and a spot shows up near the lintel of the other doorway. Between the solstices, spots of sunlight appear along the wall between the two doors. It is possible that the inhabitants marked off a crude solar calendar on the original plaster wall that has disintegrated with time. At the autumnal and vernal equinoxes, the setting Sun is aligned with the third entrance, which opens out to a cliff overlooking the canyon, and one of the other two doorways. Thus, all the doors and windows of the room seem to have been arranged in accordance with the position of the Sun. At the Cajon cluster of ruins, a room with wall slots is similar to the room at Hovenweep Castle. As with the Castle room, the Cajon room was a later addition to the main tower, which suggests that its function supplemented the primary purposes of the towers.

Another Hovenweep solstice marker is found near the Holly House group of ruins. In a canyon about 100 meters south of the ruins two huge boulders form a sort of corridor. One boulder has an overhanging ledge. The lower section of this ledge is decorated with a row of three petroglyphs. On the left are two spirals, and on the right is a set of concentric circles believed to be a symbol of the Sun. About 45 minutes after sunrise on the mornings of and near the summer solstice, a dagger of sunlight appears to the left of the leftmost spiral and gradually stretches horizontally to the right, slicing through the top of the spiral. Another dagger appears between the center spiral and the circles and extends horizontally in both directions and nearly bisects the center spiral. A third beam of light appears to the right of the circles and elongates to the left through the center of the circles. Finally, all three light beams meet to form a single shaft of light that stretches across the entire face of the boulder. The show of shards of sunlight takes place in a span of about ten minutes.

Visiting Information

Hovenweep National Monument is located in a remote area northwest of Cortez, Colorado, near the border between southeast Utah and southwest Colorado. The Visitor Center and the Square Tower Group can be accessed via paved roads from Cortez, Colorado, and from Highway 191 south of Blanding, Utah. A two-mile loop trail takes you from the Visitors Center to the Square Tower Group, which includes Hovenweep Castle. Roads to the outlying ruins, including Holly House, are dirt and are not regularly maintained. The park service recommends the use of high clearance vehicles for visiting these sites.

The monument has a ranger station, a campground, and trails, all located at the Square Tower group. There is no entrance fee. The ranger station is open year round from 8:00 A.M. to 4:30 P.M. Here, you can find a restroom, drinking water, some literature, and a few

> Website: www.nps.gov/hove
> Telephone: 970–562–4282

convenience items for sale. There is no gasoline, food, or lodging available on the grounds. The nearest lodging is about 40 miles from the monument. The campground operates on a first-come, first-served basis. The fee is $10 per night. A few of the sites can accommodate RVs.

Mesa Verde National Park, Southwestern Colorado

Mesa Verde National Park was the first national park established to preserve the works of humans rather than nature. The Ancestral Puebloan people lived here for more than 700 years, from A.D. 600 through A.D. 1300. For the

Cliff Palace at Mesa Verde National Park. The four-story square tower can be seen at the center.

first 600 years, they lived mostly on the mesa tops. Not until the last 100 years did they move into the cliff dwellings for which the park is most famous. Today, the park is home to more than 4,000 archaeological sites, including 600 cliff dwellings. Two sites, Cliff Palace in conjunction with the Sun Temple, may have served an astronomical purpose.

One of the most enigmatic ruins at Mesa Verde is called the Sun Temple. Don't let the name fool you. Although archeologists believe the structure probably served as a ceremonial center for the 600 or so nearby inhabitants, it is not at all clear whether those ceremonies had anything to do with the Sun. The ruin was given the name "Sun Temple" only because a boulder on the southwest corner of the ruin is decorated with a sun-like design. The main section of the ruin consists of a D-shaped windowless double wall that encloses two circular rooms that may at one time have been towers. On the western side of the Temple is an "annex" with two smaller circular rooms along with several small rectangular rooms. Construction of the Sun Temple began around 1275, but the building was abandoned and left uncompleted when the Mesa Verdeans suddenly left the area the next year. As explained below, the Sun Temple may have been built partly as an artificial horizon marker by sun-watchers at Cliff Palace.

Home to an estimated 100 to 120 residents occupying 217 rooms and 23 kivas, Cliff Palace is the largest cliff dwelling at Mesa Verde. On the winter solstice, the Sun, as seen from Cliff Palace, sets on the flat and featureless southwestern horizon, featureless except for the Sun Temple perched on top of a mesa nearly 300 meters across the canyon. A possible sun-watching station has been identified at the extreme southern end of Cliff Palace. Here, you find a small level platform with an eight-centimeter-diameter basin carved into the rock. If you stand directly over the basin on the winter solstice, then you see the Sun set between the two main towers of the Sun Temple. From this same vantage point, the smallest tower in the temple's western annex could have served as a horizon marker, giving a 20-day advance warning of the approaching solstice. Of course, these alignments only hold true if the circular rooms found at the Sun Temple did in fact form the base of towers that rose well above the temple's walls. In today's ruins, these circular rooms are not tall enough to serve as horizon markers; whether they ever did is anybody's guess.

In addition to solar observations, credible circumstantial evidence suggests that the Mesa Verdeans made systematic observations of the Moon. The site of these lunar observations is a small third-story window in the four-story Square Tower at Cliff Palace. From this window, on the night of the major southern standstill, the Moon can be seen setting between the two towers of the Sun Temple. The idea that this alignment is intentional rather than coincidental is supported by three nearby pictographs painted in red on the wall of the third story of the tower. The first pictograph consists of four vertical lines, each with a series of 17 to 20 short tick marks in a rakelike pattern. The average number of ticks per vertical line is between 18.5 and 18.75, which corresponds well with the 18.6-year lunar standstill cycle. The number of years it takes to complete four standstill cycles agrees with the number of years that Cliff Palace was occupied.

A second pictograph is a rectangle divided in half by a line with 12 tick marks emanating from both sides. On either side of the dividing line are 6 zigzags for a total of 12, which is significant because of the approximately 12 lunar months in a year. The zigzags could be interpreted as depicting the back-and-forth motion of the Moon along the horizon, both for the rising Moon on the left half of the diagram and the setting Moon on the right half. Directly below the rectangle pictogram is a third pictogram showing two sets of three triangles separated by a dozen little circles. The triangles might represent the sharp peaks of the La Plata Mountains, which can be seen on the northeastern horizon, and the circles could again represent the 12 lunar months.

Of course, zigzags and rows of tick marks are common decorations in Ancestral Puebloan art; alternative, nonastronomical interpretations of the pictograms may also be correct. Another observation argues against the lunar cycle interpretation of the alignments and pictograms: although modern Pueblos demonstrate an interest in the phases of the Moon for keeping calendars, they do not show a knowledge of or interest in longer lunar cycles, such as the 18.6-year standstill cycle.

Visiting Information

Mesa Verde National Park is located in southwestern Colorado, about an hour's drive east from Cortez, Colorado, on Highway 160 or an hour-and-a-half drive west from Durango, Colorado, on the same highway. The park is open year-round, but some sites, tours, and facilities are seasonal. Check the website for a schedule. The entrance fee is $15 during the summer and $10 the rest of the year. The mountain road leading into the park is steep and winding, so it will take you about an hour to drive the 15 miles from the park entrance to the Far View Visitor Center. To make the most of your visit, stop at the Visitor Center to get your bearings. The Far View Visitor Center is open from the spring through the fall from 8:00 A.M. until 5:00 P.M. The Chapin Mesa Archaeological Museum, open year round, is about six miles further into the park.

The Sun Temple is one of a dozen stops along the Mesa Top Loop Road, a six-mile driving tour open from 8 A.M. until sunset. The Cliff Palace can only be visited by taking a one-hour ranger-guided tour that includes climbing several ladders. Tickets for this tour are $3.00 per person and must be purchased in advance at the Far View Visitor Center.

The park has all the amenities you will find at any national park, including food and lodging. In addition, half-day

Website: www.nps.gov/meve
Telephone: 970–529–4465

and full-day guided bus tours are available for an additional fee.

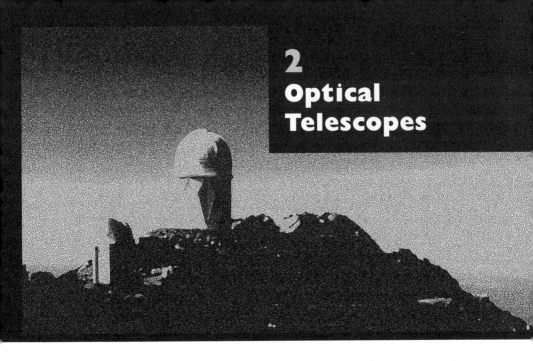

2
Optical Telescopes

O telescope, instrument of much knowledge, more precious than any scepter!
Is not he who holds thee in his hand made king and lord of the works of God?

Galileo

In 1609, Galileo Galilei, copying a design from the Dutch lens grinder Hans Lippershey, built a crude little telescope, pointed it skyward, and saw spots on the Sun, mountains on the Moon, and moons around Jupiter. Galileo was the first to use a telescope for astronomical research, and his discoveries revolutionized science. More than 300 years later, Edwin Hubble used a giant telescope on a California mountaintop to prove that we live in a universe teeming with galaxies, all hurtling away from each other as they are swept along by the expansion of space. Because it's a bit impractical to bring a star into a laboratory and perform experiments on it, astronomy is, for the most part, an observational science; you just have to sit back and watch. As the discoveries of Galileo and Hubble show, our understanding of both the universe and our place in it depends on having telescopes available to harvest visible light and other types of electromagnetic radiation coming from outer space.

Electromagnetic waves consist of vibrating electric and magnetic fields ripping through space at the incredible speed of 186,000 miles per second. But space is mind-bogglingly vast. Astronomers measure distances in space with a unit called a light-year—the distance light travels in one year. One light-year is equivalent to roughly six trillion miles. Aside from the Sun, our nearest stellar neighbor is a star appropriately named Proxima Centauri, some 4.2 light-years from Earth. This means that it takes light from Proxima Centauri 4.2 years to get to the Earth. When we look at Proxima Centauri in the night sky, we do not see it as it is today, but as it was 4.2 years ago; we are literally looking back into time 4.2 years. But Proxima Centauri is the nearest star. If we look farther into space, we see further back into time. Modern telescopes can see objects billions of light-years out into space; therefore, we can see billions of years back into time. Telescopes are, in a very real sense, time machines.

The electromagnetic spectrum is divided up into radio, microwaves, infrared, visible, ultraviolet, x-rays, and gamma rays. The radio waves are the longest waves and the gamma rays are the shortest. Of these, only visible light and radio waves can easily penetrate through the Earth's atmosphere and get to the surface. Ground-based observatories are therefore generally limited to these regions of the spectrum. This chapter describes observatories with telescopes that operate in the visible range, and chapter 3 focuses on telescopes that operate in the radio range. The other parts of the spectrum must be observed with telescopes floating in orbit above the Earth's atmosphere, which, of course, puts them well out of reach for most scientific travelers.

There are two main types of optical telescopes: refractors and reflectors. Galileo's telescope was a refractor. The first practical reflecting telescope was invented by Sir Isaac Newton in 1670. In a refracting telescope, a convex lens (thin at the edges and thick in the middle), called the objective, is positioned at the skyward end of the telescope; the lens bends or refracts the light to form an image, which is then magnified by a second lens, called the eyepiece. In a reflecting telescope, the light is collected and focused by a curved mirror called the primary. A flat secondary mirror reflects this image to an eyepiece. For reflecting telescopes, the curved mirror is usually in the shape of a paraboloid. To understand what a paraboloid is, recall that a parabola is one of those conic sections you may have studied in high school geometry class. A paraboloid is formed by rotating a parabola about an axis through its center. Variations on the reflecting telescope design include Cassegrain and Schmidt reflectors.

When it comes to telescopes, size definitely matters. The bigger the lens or mirror, the more radiation it can collect, enabling the telescope to see fainter and more distant objects. Nearly all modern telescopes used for astronomical research are reflectors. Why? Lenses can only be supported around their edges and, as you scale a lens up in size, at some point the lens begins to sag under its own weight. Make it even bigger, and the lens will break. This imposes an upper limit to the size of a lens. The largest refracting telescope ever built is located at Yerkes Observatory in Wisconsin. Its objective lens has a diameter of 40 inches and weighs about 500 pounds. This is about as big as a lens can get. It is therefore highly unlikely that the Yerkes telescope will ever relinquish its title as the world's largest refractor. The weight of a mirror, however, can be supported by its entire back side, so mirrors can be made much heavier and bigger than a lens. The largest reflecting telescope consisting of a single mirror is the 200-inch reflector at Palomar Mountain in California.

But bigger mirrors are more expensive, more difficult to manufacture, and less maneuverable. So instead of using a single large mirror, most state-of-the-art optical telescopes piece together a cluster of smaller, independently controlled, lightweight mirrors that collectively have the same light gathering capability as one large mirror. As of this writing, the largest optical telescopes in the world are the Keck I and II reflectors perched high atop the Mauna Kea volcano in Hawaii, although they will soon be eclipsed in size by a new generation of even larger telescopes. The Keck telescopes consist of 36 individual hexagonal mirrors put together in a honeycomb pattern to form a single reflecting surface 33 feet across. Even larger optical telescopes are now under construction.

Once an astronomer uses a telescope to collect the light, what does he or she actually do with it besides take pretty pictures? Astronomers use an instrument called a spectroscope to measure the wavelength of the light; they use an instrument similar to a light meter to measure the intensity of the light; and they carefully measure the direction the light is coming from. These instruments are normally attached directly to the telescope, and the data is sent to a computer for analysis. (Today's professional astronomers rarely look through a telescope by actually putting their eye to the eyepiece.) By analyzing the wavelength, intensity, and direction of the light and changes in these quantities, astronomers can deduce the physical characteristics of the object such as its mass, motion, and composition.

In the history of observatories and telescopes in the United States, two names stand out. First, the optical firm of Alvin Clark and Sons in Cambridge, Massachusetts, ground lenses for refracting telescopes during the late 1800s and early 1900s. The firm was generally regarded as the best telescope maker in the United States and maybe even the world. Many of their telescopes still bear the Clark name. Second, George Ellery Hale was an astronomer who helped found and find financing for the Yerkes Observatory, the Mount Wilson Observatory, and the Palomar Observatory. At these observatories he built the world's largest telescope not once, but four times. The 200-inch telescope at Palomar is named in his honor. Never satisfied, Hale constantly pushed for bigger and better telescopes, an obsession that drove him to the brink of madness. He invented the word "astrophysics" and founded *Astrophysical Journal,* the world's leading scientific journal in the area of astronomy. Largely as a result of Hale's efforts, the United States established itself in the twentieth century as the world's leader in astronomical research.

The first observatory built in the United States was the Hopkins Observatory at Williams College in Williamstown, Massachusetts, constructed in 1838. Today, hundreds of observatories are scattered across the United States with the largest concentration of big telescopes in the desert southwest to take advantage of the dry, stable, and clear desert air. Most large optical observatories are located on mountaintops to minimize the negative effects of the Earth's atmosphere. The observatories and telescopes themselves are often painted white to reflect light. This minimizes the heating of the materials which, in turn, minimizes the expansion of the telescope's various parts.

Given the limited number of large telescopes and the approximately 5,000 professional astronomers in North America alone, time with the telescopes is a precious commodity. Astronomers must compete for telescope time by writing proposals and submitting their ideas to observatories, where they are carefully read and judged. If the research is deemed worthy enough, the astronomers are rewarded with some coveted telescope time.

I describe ten major American observatories. These are just a few of the places astronomers go to harvest ancient starlight in their ongoing quest to decipher the secrets of the stars. I chose these observatories based on their scientific and historical importance; they are listed in chronological order from the oldest observatory to the newest. If you can't make it to these observatories, remember that most universities have an observatory, often located right on campus, and many of these have regularly scheduled observing nights for the general public. During these events, you can do what you normally can't do when you visit one of the major observatories: you can actually look through the telescope! Check with the astronomy department at your local university for details.

The U.S. Naval Observatory, Washington, D.C.

Want to know what time it is? Ask a member of this observatory's Time Service Department, and he or she can tell you the time to the nearest nanosecond. In contrast to the other observatories described in this chapter whose research efforts focus on modern astronomical topics, a main mission of the U.S. Naval Observatory (USNO) is to determine the exact time. To this end, the observatory maintains a collection of clocks that are constantly checked against each other and against the stars. These "clocks" don't look like the clocks you have at home; they are atomic clocks. The observatory's bank of atomic clocks consists of about 50 cesium-beam frequency standard clocks that keep time based on the atomic transitions of electrons in cesium atoms and a dozen hydrogen masers that keep time using the oscillations of hydrogen atoms. This ensemble of clocks forms the observatory's Master Clock and sets the standard of time for the United States and the Department of Defense.

The other major mission of the USNO is to determine the exact positions and motions of the Earth, Moon, Sun, planets, and stars: a branch of astronomy called astrometry. This data is made available to astronomers, government agencies, and to the military for navigation, precise positioning, and communications. The USNO is one of only three observatories in the world that specializes in astrometry, the others being the Royal Observatory in Greenwich, England, and the Pulkovo Observatory near St. Petersburg, Russia.

The first attempt to build a national observatory can be traced back to 1810 when a proposal was introduced into Congress. Despite support from such political luminaries as John Quincy Adams and Thomas Jefferson, legislators resisted the use of public funds to support scientific research. When the Depot of Charts and Instruments was established in 1830, the ship chronometers needed to be synchronized with astronomical events. As a result, astronomical observations began at the Depot and expanded rapidly, a development requiring the construction of permanent facilities. The needed buildings were approved in 1842, and in 1844, the Depot was reestablished as the U.S. Naval Observatory. Today, the USNO stands as one of the oldest scientific institutions in the United States.

The observatory was originally located on a hill in the "Foggy Bottom" area of Washington, D.C., at 23rd and E streets between what is now the Lincoln Memorial and the Kennedy Center. That site is now occupied by the

Navy's Bureau of Medicine and Surgery. Unfortunately, it is closed to the public, but the white dome of the original building is visible from the street. The observatory remained at this site for nearly 50 years. Under the leadership of Matthew Fontaine Maury, who has been called the father of modern oceanography, the observatory earned worldwide acclaim for advances in navigation, astronomy, and oceanography. In 1877 astronomer Asaph Hall made the most famous discovery here by identifying two tiny Martian moons, Phobos and Deimos.

As the name might suggest, Foggy Bottom was not an ideal location for an astronomical observatory. By the late 1800s, the fog problem had been compounded by smoke from nearby factories. In 1893, the observatory moved to its present location at 34th and Massachusetts Avenue, a site that at the time lay on the outskirts of the city and also commanded the highest elevation in Washington. For more than a hundred years, astronomers here have made celestial observations using the same equipment from the same site. This is significant because astronomers can discover small site- and instrument-related errors in the observations and correct for them. Today's observations can be compared with observations made decades ago, enabling observers to measure celestial positions with unrivaled precision.

The largest telescope at the USNO is the beautiful 26-inch Clark refractor. Completed in 1873 at a cost of $50,000, this was the world's largest telescope for a decade. With this telescope Hall discovered the moons of Mars. The dome features a wooden floor that can be raised or lowered to provide easy access to the eyepiece. Today, the telescope is used to make observations on double star systems and to chart the positions of the moons of the outer planets to help determine their exact orbits. This data is crucial in planning future missions to these faraway worlds.

Since 1974, the mansion at Number One Observatory Circle has served as the official residence of the vice president of the United States. As you can imagine, this, along with the fact that the observatory is a military installation, means that security is extremely tight.

Visiting Information

Free public tours of the USNO are usually offered on two Monday nights each month except during August. Due to security, you must request a reservation four to six weeks in advance. The limited number of slots available are filled on a first-come, first-served basis, so make your request as early as possible. All visitors must submit to a security screening, and personal belongings may be searched. Cameras are allowed. The 90-minute tours begin at 8:30 P.M., but you must arrive absolutely no later than 8:20 P.M. See the website for up-to-date details.

The tours include a presentation on the mission and history of the observatory, a discussion of the observatory's time-keeping responsibilities along with a peek at the Master Clock and viewing celestial objects through the 12-inch Clark refractor. If it's cloudy, you may get to see the historic 26-inch Clark refractor. This is one case where you might want to wish for clouds!

The entrance to the USNO is located off of Massachusetts Avenue at the end of

Website: www.usno.navy.mil
Telephone: 202–762–1489

Observatory Circle, NW, directly across the street from the Embassy of New Zealand. The nearest Metro stop is Woodley Park.

Lick Observatory, near San Jose, California

Owned and operated by the University of California, Lick Observatory was the world's first permanently occupied astronomical observatory built on a mountaintop. Perched at the summit of 4200-foot Mount Hamilton, the highest peak in the Diablo Mountain Range, the observatory is home to the world's second largest refracting telescope. Today, the observatory is perhaps best known as a leader in the search for planets outside our own solar system. Other areas of research include the chemical composition and evolution of stars and galaxies, brown dwarfs and other low-mass companion stars, massive intergalactic hydrogen clouds, and distant radio galaxies.

The observatory is a monument to the memory of James Lick, who, at the time of his death in 1876, was one of the richest men in California. The road to Lick's financial fortune began in South America where he was a successful piano maker. The miserly Lick saved up a tidy sum of $30,000 and moved back to the United Sates in 1848. He arrived at a little shantytown with a population of about a thousand on the coast of the new state of California. Lick predicted that the town's natural harbor would eventually turn it in to a thriving metropolis, and he used much of his savings to buy up land. The name of the town was San Francisco. Soon after Lick's arrival, gold was discovered at Sutter's Mill, and the California gold rush was on. In two short years, San Francisco's population ballooned to 20,000, and Lick made a fortune. (Interestingly enough, Lick had purchased 600 pounds of chocolate from a confectioner who had a shop in his neighborhood in Lima, Peru. He brought it with him to San Francisco and quickly sold it. He then sent word to his Peruvian neighbor, Domingo Ghirardelli, advising him to come to San Francisco and sell his chocolate.)

Toward the end of his life, Lick wanted to build a monument to honor his memory. He first considered erecting huge statues of himself and his parents, statues tall enough to be seen from the ocean. He abandoned the idea after a friend mentioned that the statues would make enticing targets should the city ever undergo a naval bombardment. Next, he thought of building a pyramid larger than the Great Pyramid in Egypt on a parcel of land he owned in downtown San Francisco. Fortunately for science, Lick's attention was diverted from pyramids to observatories. Several persons may have influenced Lick's decision. Legend has it that in 1860, Lick had met a man named George Madeira, an itinerant lecturer on astronomy who showed Lick the heavens through his small telescope. Supposedly, Madeira told Lick: "If I had your wealth . . . I would construct the largest telescope possible to construct." Physicist Joseph Henry, secretary of the Smithsonian Institution, claimed to have suggested to Lick that he follow the example of John Smithson, who had used his fortune to create the Smithsonian. But it was Lick's good friend George Davidson, an astronomer and president of the California Academy of Sciences, who was most responsible for gently steering Lick in an astronomical direction. Over the course of many visits with the aging and ailing tycoon, Davidson talked about planets and stars and telescopes. As a result of these conversations, Lick decided to designate the largest single allotment of his fortune, $700,000 out of a total fortune of $3,000,000, to the building of "a telescope superior to and more powerful than any telescope yet made . . . and also a suitable observatory connected therewith." Lick appointed a Board of Trustees to carry out his wishes.

One of the most important decisions that the board had to make concerned the location of Lick's observatory. Today, we just assume that any new observatory will be built on a mountaintop, but in the late 1800s, observatories were usually built in cities near universities. No one had ever tried to build an observatory on the top of a mountain. But cities, with their lights and smoke, were becoming increasingly inhospitable environments for telescopes. So, in spite of the daunting logistical challenges that building an observatory on top of a mountain would entail, the trustees choose Mount Hamilton as the site for the observatory.

Jeff Sprague/iStock

The 36-inch Clark Refractor at Lick Observatory.

Because not even a trail led up to the summit, the first step was to build "a first-class road to the summit." Santa Clara County eagerly agreed to undertake the road project in exchange for the prestige of being home to a prominent scientific facility. The road that leads to the summit today lays along the same path as the original road. Although the road is, by modern standards, excessively winding, remember that it was originally designed for slow-moving teams of horses, not modern automobiles.

Construction of the actual observatory began in 1880 and took eight years to complete. The steep mountaintop was leveled off by blasting away 70,000 tons of rock. A deposit of clay was found about a mile from the summit. Kilns were built and the clay was used to make bricks that were fired at the rate of 10,000 a day. The observatory's main building was to house two domes, one for the giant refractor and the other for a smaller telescope. The twin domes were to be connected by a long hallway with adjoining offices, laboratories, and a library. Meanwhile, a firm in San Francisco worked on the dome, a company in Cleveland pieced together the telescope frame and mounting, and a Paris factory formed the raw glass blanks for the lens. Alvin Clark and Sons in Boston was entrusted with the delicate job of shaping the lens.

On a cold winter's night in January 1888, the 36-inch refractor, the largest telescope in the world at the time, saw first light. It held the title of world's largest telescope until the 40-inch refractor at Yerkes Observatory was built about ten years later. Unfortunately, James Lick did not live to see his monument completed. As was his wish, he is buried at the base of his telescope. There, a bronze plaque reads simply: "Here Lies the Body of James Lick."

Today, two of the observatory's nine research-grade telescopes can be viewed by the public: the historic 36-inch Lick Refractor and the 120-inch Shane Reflector. The giant lens of the refractor is still carried on its original mounting at the end of a 57-foot-long white telescope tube that is four feet wide in the center and tapers to a diameter of about three feet at the top end. The beautiful inlaid hardwood floor of the dome can be raised or lowered through a total distance of 16.5 feet bringing the eyepiece to within easy reach of the observer. Astronomer E. E. Barnard made the first major discovery with the telescope in 1892 when he spotted a fifth moon around Jupiter, the first to be discovered since Galileo found the original four in 1610. The moon was named Amalthea and the telescope went on to discover several other moons orbiting Jupiter. The telescope has been used extensively for photographic work, for determining the orbits of binary stars, and for measuring the speed at which stars are moving directly toward or away from us (astronomers call this speed "radial velocity"). This later work made it possible for astronomers to estimate how our own solar system was moving within the galaxy. Recently, the telescope has been used to take photographs of star clusters. Astronomers compare these photographs with similar photographs taken in the 1970s in an effort to determine exactly which stars are members of the clusters.

The other telescope that can be viewed by the public is the 120-inch Shane Reflector, named in honor of Donald Shane, the observatory's director who led the effort to fund and build the telescope. This is the largest telescope at Lick Observatory and one of the larger reflectors in the world. The mirror for this telescope was originally a glass test blank for the 200-inch Palomar telescope. After Caltech sold the blank to Lick Observatory, it was hauled up Mount Hamilton to be ground and polished into a 4.5-ton mirror. Designed for maximum versatility, the telescope saw first light in 1959, at which time it was the second largest telescope in the world.

The observatory has several other telescopes of note: the 36-inch Crossley Reflector for historically important studies of stellar evolution and universal expansion; the 40-inch Anna L. Nickel Reflector to study stellar evolution, galactic structure, and quasars; and the 20-inch Carnegie double astrographic telescope to measure the motion of stars by taking simultaneous photographs, one in blue light and the other in yellow light. Of more recent vintage is the Katzman Automatic Imaging Telescope (KAIT), a 30-inch robotic telescope that is programmed to automatically search for supernovae in an effort to catch these exploding stars early in their brief lifetimes so that they may be studied more completely. Work is currently underway on a new telescope called the Automated Planet Finder Telescope (APF). This telescope will be fully automated like the KAIT but will hunt for rocky earth-like planets orbiting other stars.

Scientists now use the Anna Nickel Telescope for an optical SETI program. Some astronomers believe that the most likely method an advanced civilization would use to send a signal to its intelligent neighbors would be in the form of an intense beam of visible light, similar to a laser beam. Visible light beams have several advantages: they travel through space easily with little interference; they are unidirectional (that is, the source can be pinpointed with great precision); and, because of their high frequency, they can send vast amounts of information. Astronomers are using this telescope to look for artificially produced bursts of visible light—a signal from ET!

Visiting Information

Lick Observatory is open to visitors daily except on Thanksgiving, Christmas Eve, and Christmas. Hours are weekdays from 12:30 P.M. to 5:00 P.M. and 10:00 A.M. to 5:00 P.M. on weekends. There is no admission charge. Visitors should drive to the Main Building where they can view the exhibits and catch a free tour. Tours begin at the small gift shop on every half-hour starting at 1:00 P.M. on weekdays and 10:30 A.M. on weekends. The tour includes a fifteen-minute talk and a visit to the 36-inch refractor. Visitors may also walk over and ogle the 120-inch Shane reflector from the Visitors Gallery.

During the summer, a special Summer Visitors Program is offered on a half dozen or so evenings. The program includes a slide lecture on the history of the observatory and a presentation by a professional astronomer on their research or a topic of current interest. But the main attraction is getting the opportunity to look through the 36-inch refractor and the 40-inch reflector. Tickets must be purchased for the Summer Visitors Program, and availability is limited. See the website for details.

Website: www.ucolick.org
Telephone: 408–274–5061

"Music of the Spheres" is a summer concert series held in the Great Hall of the Main Building. The performances are followed by a talk given by an astronomer and a viewing through one of the telescopes. Tickets must be purchased for the concerts. Lick Observatory is about a one hour drive from San Jose. The mountain road is winding and narrow, so be careful!

Lowell Observatory, Flagstaff, Arizona

The famous Lowell Observatory in Flagstaff, Arizona, with its exceedingly rich and fascinating history, is forever linked with two of the worlds in our solar system: Mars and Pluto. It all started when Percival Lowell, a wealthy Bostonian (the city of Lowell, Massachusetts, bears the family name) and avid amateur astronomer, decided to spend some of his money on an observatory. Lowell had developed an intense interest in the planet Mars and wanted a place where he could go to study and observe the mysterious red planet.

Lowell was convinced that Mars was home to an ancient and advanced civilization. In the late nineteenth century, maps of Mars suggested an earthlike planet with ice caps, oceans, lakes, and continents, but the observations of Italian astronomer Giovanni Schiaparelli stirred Lowell's imagination. In 1877, a year that brought the Earth and Mars close together, Schiaparelli reported seeing lines criss-crossing the surface of Mars, features Schiaparelli called *canali,* the Italian word for "channels" or "grooves." When Schiaparelli's work was translated into English, his "channels" became "canals." Channels could be carved by natural forces, but canals were the work of intelligent beings.

Lowell wanted to see these canals. In 1893, he sent his assistant, Andrew Douglas, who later founded the science of dendrochronology or tree-ring dating, to the desert southwest to find the best location for an observatory. After visiting Tombstone, Tucson, Tempe, and Prescott, Douglas advised Lowell that the best "seeing" was in Flagstaff. The exact location was just west of town on top of a hill 330 feet above Flagstaff. The hill later became known as Mars Hill. Lowell was in a hurry because the Earth and Mars would once again be close together in 1894 so he admonished Douglas to get on with the work quickly. On May 28, 1894, the 39-year-old Lowell arrived in Flagstaff and immediately commenced his observations of Mars, which he continued for the next 22 years. The first winter in Flagstaff brought heavy snow and poor observing conditions, which caused Lowell to reconsider his selection of Flagstaff as the location for his observatory. He renewed his search for an ideal site, but Flagstaff, though not perfect, eventually won

Percival Lowell observing through the 24-inch Clark Refractor.

out. In 1896, a 24-inch refracting telescope built by Alvin Clark and Sons was delivered and became the observatory's premier telescope.

When Lowell looked at Mars through his telescopes he saw a planet covered with canals that radiated outward from central points. Lowell believed that the canals were a planetwide irrigation system, transferring scarce water from the melting polar ice caps to the dry, arid regions closer to the equator. He speculated on the Martian political system deducing that, since the canals spanned the Martian globe, the Martians must operate under a worldwide government. He even went so far as to identify a spot on the surface called Solis Lacus, the Lake of the Sun, as the probable capital city of Mars because a number of prominent canals intersected there. Lowell published his ideas in popular magazines and wrote three major books on the subject. He was an entertaining writer, and his Martian musings captured the public imagination. Part of the popular appeal may be attributable to the fact that the late nineteenth and early twentieth century was a time of giant canal building on Earth. The Suez Canal, for example, was completed in 1869 and the Panama Canal in 1914. If Earthlings could build big canals, then surely an older and technologically more advanced Martian civilization could as well.

But why didn't the canals show up in photographs of Mars? Lowell explained that the canals can only be seen when, for just a moment, the atmosphere steadies and the surface can be seen clearly. In the next instant, the atmosphere becomes turbulent again and the detail is lost (thus Lowell's insistence on a location with good seeing). Astronomical photographs involve exposing the film over a period of time. This blurs the details and renders the canals invisible. In the case of the canals, Lowell argued, the eye was superior to the camera because the eye could capture those fleeting instants in time when the surface could be seen with exceeding clarity. The human brain could register those images, and the observer could sketch them on paper.

Did other astronomers see the canals? Some did; some did not. Lowell dismissed those who did not see the canals by explaining that their observatories were not in the best locations or that they lacked observing experience or that they were biased against the canal theory. The astronomers who did see the canals drew maps that were often similar to Lowell's own maps and hundreds of the canals were given names.

Not until 1971, when the *Mariner 9* spacecraft orbited Mars and sent back close-up photographs of the surface, unobscured by the Earth's turbulent atmosphere, were Lowell's theories tested directly. The *Mariner* photographs and other images taken by subsequent spacecraft show not even a hint of canals. The robotic rovers wandering around on the surface have not fallen into a canal. There are no canals. Lowell was wrong. How do you explain his (and others') mistakes? Some have suggested a quirky failure of the human hand-eye-brain combination. Others have suggested that Lowell and his followers, enamored with the idea of a Martian civilization, saw what they wanted to see.

Although Lowell was wrong about Mars, he and the astronomers he hired turned out to be right about other important astronomical questions. The most cosmologically significant discovery ever made at Lowell Observatory was the result of work done by Vesto Melvin Slipher, an Indiana farm boy that Lowell hired as a "temporary" assistant in 1901. Slipher's first job was to learn how to use the observatory's shiny new spectrograph, an instrument that dissects light into its constituent wavelengths much like a prism disperses light into its constituent colors. In 1909, Lowell asked Slipher to do a spectrographic study of faint, fuzzy, swirling patches of light called spiral nebulae. There was a great debate raging in the astronomical community as to the nature of these nebulae. Some astronomers thought they were clouds of gas and dust within our own Milky Way galaxy, while others were convinced they were "island universes," galaxies similar to, but separate from, our own. When Slipher produced a spectrogram of one of these spiral nebula in the constellation of Virgo, he noticed the spectral lines were shifted toward the red end of the spectrum. This meant it was moving away from the Earth. The speed can be calculated from how big the shift is: the bigger the shift, the faster the object is moving. Slipher calculated that the Virgo spiral nebula was moving at the fantastic speed of two million miles per hour. Slipher measured the velocities of fifteen spiral nebula, nearly all of which exhibited a redshift, and presented his findings to a meeting of the American Astronomical Society in 1914. In his speech, Slipher made a prescient remark when he noted that the red shifts indicated a "general fleeing from us or the Milky Way." In the 1920s, Edwin Hubble, building on Slipher's discovery, measured the distance to the fleeing nebula and proved that we live in an expanding universe. Upon Percival Lowell's death, Slipher became the observatory's second director.

In spite of the importance of Slipher's discovery, the Lowell Observatory is best known for finding that little world on the outskirts of the solar system, the world that has stirred up so much controversy of late—Pluto. Based on tiny deviations in the expected orbit of Uranus, Percival Lowell predicted the existence of an unknown Planet X in 1902. Lowell employed a dozen mathematicians who worked for ten years calculating the required location for Planet X to account for these orbital perturbations. Lowell himself conducted several unsuccessful searches for Planet X between 1905 and 1915. In 1929, Slipher hired a 22-two-year-old amateur astronomer from Kansas named Clyde Tombaugh as an assistant to help look for Planet X. Tombaugh's duties included stoking the observatory's furnaces, leading afternoon tours, and preparing a thirteen-inch refracting telescope for the planet hunt. Tombaugh's method was to take a photograph of an area of the sky that might hold the elusive planet and then take another photograph of the same area of the sky several nights later. Tombaugh compared the two photographs by using an instrument called a blink comparator, basically a microscope through which two photographs could be viewed, or "blinked," in rapid succession. If a relatively nearby object moved against the motionless background of the distant stars, then the observer would see a streak of light.

On the night of January 23, 1930, Tombaugh took a photograph of a piece of the sky in the constellation of Gemini. On the night of January 29, he took a second photograph of the same piece of sky. Three weeks later, he examined the photographs with his blink comparator. At 4:00 P.M. on February 18, Tombaugh saw a little speck of light "popping in and out of the background" and realized he had found Planet X. The ever cautious Slipher had the object tracked and photographed for three more weeks just to make sure it was a planet. The announcement of the discovery of the ninth planet in our solar system was made on March 13, 1930, the seventy-fifth anniversary of Percival Lowell's birth.

Suggestions as to the name of the new planet began to pour in from all over the world. An eleven-year-old English schoolgirl sent in the name Pluto, after the Roman god of the underworld. The name seemed appropriate for an object out in the far reaches of the solar system and besides, the first two letters of Pluto were Percival Lowell's initials, a subtle, but fitting tribute to the astronomer who predicted the little planet's existence. On May 1, 1930, Slipher officially proposed that the new planet be named Pluto. Pluto's tenure as a planet lasted for 76 years. In 2006, the International Astronomical Union demoted Pluto to the status of "dwarf planet."

As the city of Flagstaff grew, so did the light pollution at the Mars Hill site. In the 1960s, Anderson Mesa, a location 12 miles southeast of the city, was chosen for future telescopes. Although it is not open to the public, Anderson Mesa houses most of Lowell Observatory's current research. The largest telescope at Anderson Mesa is the 72-inch Perkins telescope, named in honor of Hiram Mills Perkins, an Ohio hog farmer and Ohio Wesleyan faculty member who founded the Perkins Observatory in Ohio in 1924. Perkins contributed nearly $200,000 toward the building of this telescope. It was moved from Ohio to Flagstaff in 1961.

Other telescopes on the mesa include the 42-inch John Scoville Hall telescope used mainly for stellar research (the telescope is named after the observatory's fourth director, who acquired it); a 31-inch reflector that was originally used by the U.S. Geological Survey for mapping the surface of the Moon is now part of the National Undergraduate Research Observatory and is used by college students to hone their research skills; a small, fully automated Planet Search Survey telescope (PSST) was designed to detect giant extrasolar planets; and a refurbished 24-inch Schmidt telescope is dedicated to the Lowell Observatory Near Earth Object Search (LONEOS Project). The Schmidt telescope's wide field of view allows it to see large swaths of sky in search of asteroids or comets that may be on a collision course with the Earth.

Anderson Mesa is also home to the Navy Prototype Optical Interferometer (NPOI), a joint project between Lowell Observatory and the U.S. Naval Observatory. Completed in 1994, this interferometer consists of mirrors arrayed in a Y-shaped pattern covering an area of 15 acres. This arrangement enables NPOI to act as if it were one giant telescope 1,475 feet in diameter. NPOI allows astronomers to determine the positions of stars with incredible accuracy and to measure stellar properties such as size and changes in the surface.

The observatory's latest project is a joint effort with Discovery Communications, Inc. to build a giant 4.2-meter telescope at a location known as Happy Jack, 40 miles southeast of Flagstaff. With an expected completion date of 2009, the Discovery Channel Telescope (DCT) will search for extrasolar planets, explore the Kuiper Belt, and hunt for those pesky Near Earth Asteroids.

Visiting Information

You can visit the Lowell Observatory during the daytime, when you catch a guided tour of the grounds, or at night, when they actually let you look through some of the telescopes. Of course, the true scientific traveler will want to do both! With a little luck, you may get to gaze at Mars through the 24-incher just like Lowell did—but don't expect to see any canals.

Day or night, your visit begins at the Steele Visitor Center. The main exhibit here is called "Tools of the Astronomer," and, as you browse through the displays, you'll learn about the scientific instruments astronomers have at their disposal: telescopes, photometers, spectroscopes, lenses, prisms, cameras, and computers. Not to be missed is Slipher's 60-foot slide rule. For the younger readers out there—slide rules were calculating instruments used before the advent of electronic calculators. The accuracy of a slide rule was proportional to its length. The standard slide rule was 10 inches long, but Slipher evidently required something a little more accurate.

You are free to wander the grounds of the observatory, but the buildings are locked, so you will definitely want to catch a guided tour to see inside. The guided tours start at 10:00 A.M. and are offered every hour on the hour until 4:00 P.M. The tours last 45 minutes and include four stops: the 24-inch Clark refractor, the Pluto discovery telescope, the Pluto walk, and the Slipher Building. The first stop is usually the dome that houses the famous 24-inch Clark Refractor. The 40-foot wide wooden dome is in the shape of a truncated cone and revolves on car wheels, most of which come complete with hubcaps. The shutter opening was originally covered by a canvas curtain. The 30-ton telescope has a 32-foot long tube and cost Lowell $200,000 in 1896. This is the telescope that Lowell used to study Mars and that Slipher used to discover the red-shift. Later, the telescope was used by Slipher's brother who, over the course of five decades, took thousands of astronomical photographs on glass plates that are still kept in the observatory's archives. More recently, the telescope has also been used to study double star systems. It is fitting that Lowell is entombed in the elaborate mausoleum next to his 24-inch telescope. His wife thought it was too gaudy and refused to visit it after he was buried there.

Next up is the Pluto Dome, home of the Pluto Discovery Telescope. The 13-inch refracting telescope was purchased with a gift of $10,000 by Lawrence Lowell, then president of Harvard University. This is the actual telescope that Clyde Tombaugh used to discover Pluto. The Pluto Walk is a 350-foot-long scale model of the solar system that illustrates the relative sizes and distances of the planets, asteroids, comets, and Kuiper Belt objects in relation to the Sun. Blue signposts along the walk provide information about each planet and brass markers embedded in the walkway show the maximum deviation from the average distances.

Your tour ends in the historic old library inside the rotunda of the Slipher Building, built in 1916. Here, you see the very first Lowell Observatory telescope: the original Clark refractor that Douglas took with him on his 1893 quest to find a location for the observatory. He used the telescope to test the skies for good seeing. The blink comparator that Clyde Tombaugh used to discover Pluto is on display, as is the restored spectrograph used by V. M. Slipher to discover the redshift of the galaxies. Six Mars globes hand drawn by Lowell and profusely decorated with canals are on display along with some of his personal items, including his typewriter, awards, and photographs. There is information describing the observatory's role in the Apollo Moon missions along with a guest book signed by Neil Armstrong and a few of his fellow astronauts. Stand back and take in the elegant late-nineteenth-century architecture of the room highlighted by the graceful spiral staircase that leads to the shelves of books. Don't miss the lamp in the shape of Saturn, complete with rings, hanging from the ceiling.

On a nighttime visit, when the skies are clear, you have a chance to peer through the famous Clark refractor or the 16-inch McAllister reflector and see a planet, a nebula, or a star cluster. The observatory hosts special viewing events if something unusual, like a comet or an eclipse, is occurring. On cloudy nights, a multimedia presentation on the night sky is offered, along with a visit to the Clark telescope.

From March through October, the Visitor Center is open from 9:00 A.M. to 5:00 P.M. The rest of the year, the center opens at noon. For nighttime visits, the center opens at 5:30 P.M., and programs are offered Monday through Saturday from June through August and on Wednesdays, Fridays, and Saturdays the rest of the year. Admission is $6 for adults, $5 for AAA members, seniors, and college students, $3 for youth ages 5 to 17, and children under four are free. There is a separate admission fee for day

> Website: www.lowell.edu
> Telephone: 928–774–3358

and night programs. Night tickets can be purchased either during the day or when the center opens in the evening. Lowell Observatory is located at 1400 West Mars Hill Road in Flagstaff, Arizona. See the website for directions.

Yerkes Observatory, Williams Bay, Wisconsin

They just don't build observatories like this anymore! In contrast to the functional but bland buildings that hold modern telescopes, the beautiful, century-old gothic building that is home to the world's largest refracting telescope is rich in architectural detail. How did a place like this come to be? In the late 1890s, the first president of the newly established University of Chicago, William Rainey Harper, wanted to expand the university's curriculum into the natural sciences. In pursuit of this vision, Harper convinced young George Ellery Hale to be employed as an assistant professor. As part of the deal, Harper agreed to build a large observatory for not less than $250,000 within two years. Hale soon learned of the existence of two 42-inch disks of glass that were gathering dust in the shop of the famous optician Alvin Clark. The disks had been ordered by the University of Southern California, but the funds to pay for the disks had not materialized. Hale coveted the disks and convinced Harper to buy them and use them to build the largest telescope in the world. Harper saw this as an opportunity to attract scientific talent to the fledgling university and agreed. Harper and Hale convinced Chicago businessman Charles T. Yerkes, who had made a fortune financing Chicago's elevated electric railway system, to fund the building of the observatory.

The trio agreed that the observatory would be built within a hundred miles of Chicago, although building the observatory in Chicago itself was ruled out not because of the city lights, but because of pollution from the smoke and soot produced by coal-burning steam engines that powered the trains and local industry. The site at Williams Bay on Lake Geneva was cho-

The dome of the 40-inch Clark Refractor at Yerkes Observatory in Wisconsin.

sen because it was a popular summer resort area with regular train service to and from Chicago, future industrial development was unlikely, and the land was available as a donation.

The highlight of a visit here is walking into the dome and seeing the giant 40-inch refractor. At the time it was built, it was the largest telescope of any kind in the world and is still, to this day, the largest refracting telescope in the world. The telescope is 63 feet long, weighs more than nine tons, and yet is so precisely balanced that it can easily be moved with one hand. Rather than use a ladder to get to the eyepiece of the telescope, the entire 73-foot diameter wooden floor acts as an elevator suspended in air by cables and counterweights. As demonstrated on the tour, at the flip of a switch, a motor lifts or lowers the floor to the desired position. The operation of the 90-foot dome, one of the largest ever built, is also demonstrated on the tour. The inertia of the rotating dome carries it past the point where the motor is switched off so positioning the dome takes some practice.

These names of the astronomers who have worked and studied here read like a "who's who" of late nineteenth- and twentieth-century astronomy: E. E. Barnard, Gerard Kuiper, Otto Struve, Bengt Strömgren, Edwin Hubble, and Subrahmanyan Chandrasekar. Upon coming to America shortly after winning the 1921 Nobel Prize in physics, Albert Einstein is quoted as saying that he'd rather visit Yerkes Observatory than Niagara Falls; indeed, he did visit the observatory. As a graduate student at the University of Chicago, Carl Sagan spent a year in residence here. Sagan and other staffers gave public lectures on Saturday afternoons. After one of these talks, Sagan mistakenly left the clock drive on, and the telescope was found some hours later tipped at a precarious angle, thereby producing one of the gouges in the wooden floor.

Visiting Information

Yerkes Observatory is located in Williams Bay, Wisconsin, about an hour's drive from Chicago's northern suburbs. Free public tours are offered every Saturday morning throughout the year at 10:15 A.M., 11:15 A.M., and 12:15 P.M. A voluntary donation of $5 is suggested. The tour begins with an interesting half-hour talk on the history and architecture of the observatory. Then it's on to see the famous refractor. After the tour, you can read about the research being done at Yerkes on the bulletin boards along the hallway. At the end of the hallway displays of a few old astronomical research instruments include an ultraviolet camera and a spectrograph, which was used to map the spiral structure of the Milky Way galaxy for the first time. The small Quester Museum documents the lives of some famous astronomers who worked and lived here. There is a small gift shop in the main lobby with post-

> Website: http://astro.uchicago.edu/yerkes/
> Telephone: 262–245–9805

cards, T-shirts, sweatshirts, and astronomy-related items. There are a couple of small motels in Williams Bay, but plenty of lodging is available in the picturesque resort town of Lake Geneva, only six miles away.

Mount Wilson Observatory, Pasadena, California

Discoveries made at the Mount Wilson Observatory dominated astronomy in the first half of the twentieth century. In 1908, George Hale discovered that sunspots are caused by the Sun's magnetic field. In 1917, Harlow Shapley figured out where our solar system is located in relation to the Milky Way galaxy. In 1919, Albert Michelson used an interferometer to make the first accurate determination of the size of a star. And of course, most famously, in the 1920s, Edwin Hubble discovered that our Milky Way galaxy is but one of billions of galaxies in the universe. Hubble and his associate,

Milton Humason, then found that nearly all of these galaxies are flying away from our own galaxy at a speed that is directly proportional to its distance. These twin discoveries led to the conclusion that we live in an expanding universe, a universe that started with a bang about 13 billion years ago. In light of all these discoveries, a strong case can be made for the claim that Mount Wilson is the most scientifically productive astronomical observatory in history.

The observatory got its start when George Ellery Hale visited Mount Wilson in 1903 and was overwhelmed by the excellent observing conditions. The air above Mount Wilson is steady with very little turbulence. In fact, at an altitude of 5,715 feet, Mount Wilson has the best "seeing" of any spot in North America. Hale had just built the 40-inch refractor at Yerkes Observatory in Wisconsin and was looking for a place to build an even bigger telescope. In 1904, he obtained funding from the Carnegie Institution, and work on the observatory began. Hale had a particular interest in the Sun because it was, after all, the nearest star and could be studied in much greater detail. By understanding the Sun, we could understand the stars. The first permanent instrument on the mountain was the Snow Solar Telescope, a special telescope that lies horizontally to observe the Sun. This telescope produced the best solar images and spectrographs available at the time, but it was not good enough for Hale.

The problem with a horizontal solar telescope is that the observations are, of necessity, made during the day when the ground is rapidly absorbing and reemitting radiation into the atmosphere. The heat rising from the ground distorts the images of the Sun. (At night, ground radiation is at a minimum so traditional telescopes do not have to deal with this problem.) To remedy this annoyance, Hale decided to place the light-collecting mirrors at the top of a tall tower. The mirrors reflected a beam of sunlight down through a focusing lens. The resulting image is then directed into an observing laboratory. The height of the tower not only solves the heating problem, but it also results in a larger image of the Sun and a more widely spread out spectrum. In 1908, the world's first solar tower was completed at Mount Wilson. Using this telescope, Hale observed swirl-like features around groups of sunspots. The swirls reminded Hale of iron filings sprinkled on the poles of a magnet, and he noticed that the swirls were in opposite directions around adjacent sunspots, which suggested north and south poles. On June 25, 1908, Hale took a spectrograph of a sunspot that showed the Zeeman Effect, the broadening or splitting of a spectral line caused by a magnetic field. This proved a connection between sunspots and magnetism and detected the first magnetic field beyond the Earth.

A second major discovery made with the 60-foot tower came about in the 1960s when physicist Robert Leighton noticed features on the surface of the Sun that were oscillating with a period of about five minutes. These solar oscillations were dubbed solar quakes, and the new field of helioseismology, the study of the interior of the Sun, was born. Today, the 60-foot solar tower, operated by the University of Southern California, is still used for helioseismological research.

Once Hale had discovered magnetic fields on the Sun, he wanted to study them in more detail. This, of course, required an even taller tower so that the spectrum could be spread out even more, revealing even finer detail. So, a 150-foot tall solar tower was completed in 1912. This tower actually consists of two towers: the inner, solid tower rises straight up and supports the optics at the top; the framework of the outer tower supports the dome. This arrangement allows the structure to shake with the wind without vibrating the sensitive optics. The 150-foot solar tower remained the largest solar telescope in the world until 1962, when it was surpassed by the McMath-Pierce Solar Telescope at Kitt Peak. Much solar research has been conducted here, including daily (since 1917) meticulous hand drawings of sunspots and their accompanying magnetic fields. UCLA now operates the tower, which is used primarily to study magnetic fields on the surface of the Sun.

Mount Wilson's stable air makes it a particularly good location for interferometry, a technique where the light from two or more telescopes is combined to produce a more detailed image. Working at Mount Wilson, the Nobel Prize–winning physicist Albert Michelson pioneered the use of interferometers for astronomical research. Michelson devised a technique that enabled him to accurately measure, for the first time, the diameter of a star. The first star to have its girth measured was the red giant star Betelgeuse, a bright star in the constellation of Orion. It turned out that Betelgeuse has a diameter about 600 times the diameter of the Sun. If placed at the center of our solar system, Betelgeuse would reach more than halfway to Jupiter, engulfing the orbits of Mercury, Venus, and the Earth. Michelson's original 20-foot stellar interferometer, though no longer in use, is on display in the CHARA exhibit room.

Mount Wilson is home to two interferometers in use today. The Berkeley Infrared Spatial Interferometer (ISI) consists of an array of three telescopes, each of which is mounted in a trailer so they can be repositioned. Installed in 1988, the ISI has been used to measure the diameters of stars and the temperature, density, and composition of matter surrounding stars. Operated by the University of California at Berkeley, the ISI was built under the direction of Charles Townes, who won the Nobel Prize in physics for coinventing the laser.

Mount Wilson's other interferometer is operated by the Center for High Angular Resolution Astronomy (CHARA) at Georgia State University. The CHARA Array consists of six telescopes arranged in a Y formation with two telescopes forming each branch of the Y. The domes of the two telescopes forming the south arm can be seen near the dome of the 60-inch reflector. Visible nearby are the pipes that take the starlight to the beam-combining building located near the 100-inch dome. The CHARA Array is the largest interferometer in the world operating at visible wavelengths.

But Mount Wilson is most famous for its optical telescopes. This story starts back in 1896 when Hale received a 60-inch wide glass disk as a gift from his father. (What better gift for a man with a passion for telescopes?) Hale used the blank to form a mirror for a reflecting telescope. Teams of mules hauled the parts for the telescope up to the mountaintop along a winding, narrow dirt road. Part of the original road can still be seen directly south of the observatory cutting across Mount Harvard. The 60-inch Hale Reflector saw first light on December 8, 1908, and claimed the title of world's largest telescope. Among its many accomplishments, the telescope was used to determine the size of the Milky Way galaxy and to locate our position in it. Today, it is used mainly for special public outreach programs.

Before the 60-inch telescope was complete, Hale was planning an even larger telescope. Financing for the project came from John D. Hooker, a wealthy Los Angeles businessman who wanted his name on the world's largest telescope. A glass disk was ordered from the same French company that had provided the glass for the 60-inch reflector. When the disk arrived from Paris, it was full of air bubbles that astronomers believed would cause an uneven expansion of the mirror; this, in turn, would create an out-of-focus image. The company tried again and again to make a satisfactory disk, but none were deemed to be of telescopic quality. With the outbreak of World War I, no one had time to make glass disks and Hale decided to take another look at the first disk. He concluded that the bubbles might not be close enough to the surface to affect the expansion. Testing seemed to confirm this conclusion, and Hale turned the disk over to astronomer

The 100-inch Hooker Reflector at Mount Wilson.

and telescope maker George W. Ritchey, who remained skeptical about the suitability of the disk. After five years of grinding and polishing to turn the disk into a mirror, the mirror was at last installed in the telescope's frame in 1917. For a few tense hours observers thought that Ritchey might be right, but, once the mirror reached a uniform temperature, it worked beautifully. Thus, the Hooker 100-inch Reflector became the largest telescope in the world, the telescope that would be used to discover the expanding universe.

Edwin Hubble came to Mount Wilson in 1919 after earning a doctorate in astronomy from the University of Chicago, where he had become skillful in observational astronomy at Yerkes Observatory. Hubble's dissertation, titled "Photographic Investigations of Faint Nebulae," focused on the celestial objects that would eventual bring him fame. In those days, a heated scientific debate raged about the nature of these nebulae-fuzzy patches of light in the night sky. Some thought the nebulae were merely clusters of stars or clouds of gas within our own Milky Way galaxy. Others entertained the more radical notion that the nebulae might actually be independent galaxies beyond our own. Was our Milky Way galaxy the only galaxy in the universe, or were there others? The question could be answered definitively by measuring the distance to the nebulae. The Milky Way galaxy had been measured to be about 100,000 light-years across. If the distance to one of these nebulae was measured to be less than 100,000 light-years, then it most probably resided within our galaxy. But if the distance was determined to be greater than 100,000 light-years, then the nebula lay outside the borders of our own galaxy and was a galaxy in its own right.

But how do you determine the distances to objects in space? Astronomers perform this measurement by using a type of star called a Cepheid Variable. The name derives from the fact that these stars literally vary in brightness with a regular period (usually measured in days or weeks) and the first star of this type was discovered in the constellation Cepheus. There is a relationship between the period of variation of these stars and their absolute luminosity—that is, how bright the star really is in absolute terms when compared to other stars. When we look at stars in the night, we observe that some stars appear bright and some dim. But this observation could be deceptive: what appears to be a bright star might be bright only because it is closer to us than other stars; likewise, a dim star might appear dim only because it is further away than other stars. How bright a star appears to be is called its apparent luminosity. If you know both the absolute luminosity and the apparent luminosity of a star, you can calculate its distance. Astronomers refer to the Cepheid Variables as a "standard candle."

On October 4, 1923, Hubble took full advantage of the largest telescope in the world and was able to pick out, for the first time, individual stars in a nebula located in the constellation of Andromeda. He thought one star might be a nova—an exploding star—and so to track its progress he took another photograph the next night. Upon examining the photographs he realized that the star was not a nova but instead a Cepheid Variable. In his excitement, he wrote on the glass photographic plate "VAR!" for variable star. Hubble realized that he now had a way of determining the distance to the Andromeda nebula. When he plugged the numbers into the equation, he calculated the distance to be about one million light-years, a distance that placed it well outside the borders of our own Milky Way. With this discovery, Hubble ended the debate. We are but one of many galaxies. The universe was far vaster than anyone had previously imagined.

This single discovery would have assured Hubble a prominent place in scientific history. But he wasn't finished; he made a second, even more remarkable discovery. Hubble and his colleague, Milton Humason, who had once driven mule teams that had hauled equipment to the top of Mount Wilson, measured the speed of the galaxies. To measure the speed of an object, astronomers use a physical phenomenon known as the Doppler Effect. A familiar example of the Doppler Effect is the change in the pitch of the sound made by a race car. When the car is approaching the listener, the pitch of the sound is higher; as the car passes and moves away from the listener, the pitch of the sound is lower. This produces the familiar "eeeEEEOOOooo" noise at the racetrack. As the sound waves are emitted by the approaching car, the waves ahead of the car get scrunched together because the car is moving with those waves. This results in a shorter wavelength and a higher frequency or pitch. As the car moves away, the waves behind it are stretched out because the car is moving in a direction opposite to those waves. This results in a longer wavelength and a lower frequency or pitch. What happens for sound waves also happens for light waves. If a source of light is moving toward an observer, the light is shifted toward shorter wavelengths. This corresponds to the blue end of the visible part of the spectrum so we say that the light is "blue shifted." If a source of light is moving away from an observer, the light is shifted toward longer wavelengths. This corresponds to the red end of the visible part of the spectrum so we say that the light is "red shifted." The faster the object is moving toward or away from the observer, the bigger the Doppler shift. A policeman can measure the speed of a car by aiming a radar beam at it and measuring the amount of the shift. Similarly, the speed of a galaxy can be determined by the Doppler Shift in the spectra; the greater the shift, the greater the speed.

When Hubble and Humason analyzed the light from the galaxies, the spectra showed the light was shifted toward the red end of the spectrum; thus, all the galaxies were moving away from us. Now the astronomers combined their speed data with their distance data. In 1929, they discovered that the further away a galaxy was, the faster it was moving away from us. The "recessional velocity" is directly proportional to distance: a galaxy twice as far away is moving away from us twice as fast. The relationship, known as Hubble's Law, means that we live in an expanding universe. But if the universe is expanding, then just a minute ago, it was a little smaller, and a minute before that, it was smaller still. If we continue moving backward in time, the universe gets smaller and smaller. This implies that, at some time in the distant past, the universe was shrunk to a single point. This point exploded in a "big bang," and the universe has been expanding ever since. The current estimate is that this "big bang" happened about 13.7 billion years ago.

The Hooker 100-inch telescope that Hubble and Humason used to make these discoveries has heavy, industrial-age look to it with all of its steel trusses, girders, and rivets. Despite being nearly a hundred years old, the telescope's productive scientific lifetime has been extended by adding state-of-the-art instrument packages. Currently, scientists use three main instrument packages with the telescope. Two of these are "adaptive optics systems" that compensate for the already low amount of atmospheric turbulence. This results in extremely high resolution images that can sometimes even rival those taken with the Hubble Space Telescope. The third package, a fiber-optics spectrometer, can measure radial velocities with very high precision. This is currently being used in the search for planets orbiting other stars and in studies of the motion of gases within stars.

In 1984, the Carnegie Institute announced plans to close the Mount Wilson observatory to focus its attention and resources on its telescopes in Chile. Supporters formed the Mount Wilson Institute in an effort to keep the observatory running. Since 1986, the observatory has been operated by the Mount Wilson Institute under an agreement with the Carnegie Institute. The Mount Wilson Observatory Association is a volunteer group that coordinates the public outreach programs of the observatory.

Visiting Information

The observatory is open free of charge to the public from 10:00 A.M. until 4:00 P.M. every day from April 1 through November 30. A self-guided walking tour brochure and map are available on the website and at the Astronomical Museum. The tour walks you past all of these instruments. Except for the photographs encircling the room, the Astronomical Museum is sparsely appointed with only four exhibits. A set of wheels that the dome rotates on is on display along with a disk that was used to polish the telescope mirror and a scale model of the observatory. Also shown is a fly-ball governor that was used in the clockwork drive of one telescope. As you approach the footbridge that leads to the dome of the 100-inch telescope, you see a small white building on your right. This building was used as a galley where astronomers would cook their meals in the middle of the night or drink tea or coffee while waiting for clouds to pass by. Hale had strict rules against eating in the domes because of the fire hazard posed by cooking. The only telescope you can actually see is the 100-inch Hooker Reflector, which can be viewed from behind a glass window. The chair that Hubble used while making his observations sits on a platform up and to the left. Guided two-hour walking tours are conducted at 1:00 P.M. on Saturdays, Sundays, and holidays. These tours begin at the pavilion above the large parking lot and take visitors to the Astronomical Museum, past the 150-foot Solar Tower and the dome of the 60-inch reflector, and into the visitor's gallery of the 100-inch telescope.

Special tours of the grounds, including visits inside the facilities, may be arranged in advance. Groups or individuals may request a special observing session using the 60-inch reflector. These half-night or whole-night sessions require a substantial fee and must be arranged well in advance. A two-week summer school at Mount Wilson gives amateur astronomers an opportunity to learn about astronomical techniques from an expert. See the observatory association website for details.

The Mount Wilson Observatory is located high in the San Gabriel Mountains above Pasadena, California. Because the observatory is located in the Angeles National Forest, all cars are supposed to display a forest service Adventure Pass. See the website for details.

> Websites: www.mtwilson.edu
> www.mwoa.org
> Telephone: 626–440–9016

McDonald Observatory, Fort Davis, Texas

The McDonald Observatory, operated by the University of Texas, occupies the summits of Mount Locke and Mount Fowlkes in the Davis Mountains of West Texas. The area is sparsely populated and the nearest large city, El Paso, is 170 miles to the west. The isolation keeps light pollution to a minimum resulting in some of the darkest night skies in the continental United States. The observatory boasts three large telescopes: the 82-inch Otto Struve Telescope, the Harlan J. Smith 107-inch Telescope, and the giant 433-inch Hobby-Eberly Telescope.

The observatory got its start in 1932 when William Johnson McDonald, a Paris, Texas, banker with a passion for astronomy, bequeathed the sum of $800,000 to the University of Texas "to build an observatory and promote the study of astronomy." At the time, the university did not have much of an astronomy program, so the Texans enlisted the help of astronomers at the University of Chicago and Yerkes Observatory to get things up and running.

The first telescope on the mountain was an 82-inch reflector now known as the Otto Struve Telescope, named in honor of the observatory's first director. At the time of its completion in 1939, the telescope was the second largest telescope in the world after the 100-inch telescope at Mount Wilson. In the 1970s, a deranged visitor shot at the telescope with a pistol and knocked out a few chunks of glass from the mirror. Fortunately, it was not a mortal wound. The light collecting capacity was only slightly reduced, and the telescope was back in use the very next night. The top half of the telescope's 27-foot-long tube has an open structure, and the bottom half is enclosed. The tube itself is supported by a heavy steel mount. These features give the telescope a rather distinctive look, and many telescope aficionados consider it to be the observatory's most beautiful telescope. The 82-inch telescope has been continually updated and state-of-the-art equipment has been added; thus, the telescope is in demand even today. Currently, its uses include hunting for planets circling other stars.

In the early 1960s, the space age got off the ground, and NASA began planning unmanned missions to the planets. Before spacecraft explored the solar system, however, advanced telescopic observations were necessary to pave the way. In 1964, NASA and the University of Texas signed a joint contract to build a $5-million telescope to be used mainly for the study of planetary atmospheres. The Harlan J. Smith Telescope, armed with a 107-inch mirror made from fused silica, was dedicated in 1969. At the time, it was the third largest telescope in the world.

The largest telescope at McDonald Observatory and currently the fourth largest optical telescope in the world is the 433-inch Hobby-Eberly Telescope (HET). Dedicated in 1997, the HET was a joint project of Texas, Penn State, Stanford, and two German universities. The telescope was specifically designed to do spectroscopy, but at a bargain basement price of only 15 to 20 percent of the cost of other telescopes of comparable size. This savings was achieved by having the telescope always tilted at a fixed angle above the horizon. This differs from most telescopes, which can tilted at different angles. The HET sees different parts of the sky by use of a tracker mounted above the telescope. The tracker can move in six different directions allowing the telescope to view 70 percent of the night sky. The telescope's mirror consists of 91 individual hexagonal mirrors pieced together in a honeycomblike pattern. To form a near perfect reflecting surface, the mirrors must be aligned with exquisite precision. This is accomplished with a laser system inside the mushroom-shaped tower next to the HET dome. To regulate the temperature inside the dome, the exterior walls are designed like a giant set of Venetian blinds. The louvers can be opened allowing air to circulate inside the dome.

In contrast to the HET, the smallest research telescope at McDonald Observatory is the 30-inch telescope. In a real sense, this telescope is the offspring of the Harlan-Smith Telescope. When the big telescope was completed, a large hole had to be cut from the center of the 107-inch mirror to allow the light to get to the instruments. The observatory wisely decided not to let that glass go to waste and used the glass from the hole to form two 30-inch mirrors. One of those mirrors was used for this telescope whose wide field of view makes it ideal for search and survey projects.

The other 30-inch mirror was used for the Lunar Laser Ranging Station. With this 30-inch telescope scientists shoot a laser beam at reflectors left on the Moon by the Apollo astronauts. By precisely measuring the time it takes the beam to travel to the Moon and back, the distance to the Moon can be calculated to within one inch. In this way, astronomers can track the motion of the Moon and orbiting satellites.

Visiting Information

The McDonald Observatory is one of the most visitor-friendly scientific research facilities in the country. This is no accident. In his will, McDonald stipulated that the observatory be used for not only the study of astronomical science but also its promotion. As early as 1934, astronomers here were conducting public programs with a 12-inch telescope. Public outreach continues to be a primary mission of the observatory, and a wide range of activities are available for visitors. At the Visitor Center, you can explore the "Decoding Starlight" exhibit and watch multimedia presentations covering a wide range of astronomical topics. During the day, the center offers a solar viewing program where you can examine a live image of the Sun and see sunspots, flares, and prominences. Guided tours of some of the large research telescopes are offered daily at 11:00 A.M. and 2:00 P.M. A pass for these 90-minute tours may be purchased at the Visitors Center on the day of the tour. No reservations are accepted, and availability is limited so get your pass early. The passes cost $8 for adults, $7 for children ages 6 to 12, and $30 for a family of five or more. You can take a self-guided tour of the HET free of charge from 10:00 A.M. to 5:00 P.M. daily.

At night, you can take part in a "Star Party," where you can view celestial objects through the 16-inch and 22-inch telescopes at the Visitor Center's Public Observatory. Remember that moonlight makes it harder to see very faint objects so plan your visit accordingly. Dress warmly for the party, even in the summer. Star Parties are held throughout the year on Tuesday, Friday, and Saturday nights. Check the website for times. Passes for the Star Party cost $10 for adults, $8 for children, and $40 for families. Combination Passes that include both a daytime tour and a Star Party are $15 for adults, $12 for children, and $60 for families.

Before the Star Party gets started, there is a Twilight Program, a 60- to 70-minute learning experience focusing on some aspect of astronomy. This program is recommended for all ages; however, it is separate from the Star Parties and requires an additional charge.

Finally, the observatory offers special "Dinner with a Viewing" nights, which give you a rare opportunity to look through a really large telescope. If you are mainly interested in getting a spectacular view of nine or ten celestial objects, the Dinner with a Viewing on the 82-inch Otto Struve Telescope may be best for you. The Dinner and a Viewing on the 107-inch Harlan Smith Telescope is more technical. It includes a demonstration of spectroscopy, and a professional astronomer is invited to talk about their research and

> Website: http://mcdonaldobservatory.org
> Telephone: 877–984–7827

answer questions. In this program, two objects are viewed through the telescope. These special events last from four to five hours. Check the website for a schedule and prices.

Palomar Observatory, San Diego County, California

The 200-inch telescope atop Palomar Mountain is perhaps the most famous telescope in the world and has seen objects as far as 11 billion light-years away. (The location of the observatory is often incorrectly referred to as Mount Palomar. The name "Palomar" translates to "Pigeon Haven" in the local La Jolla Indian dialect.) From 1949 until 1975, it was the largest telescope in the world. The effort to build the giant telescope began in 1928 when famed telescope maker George Hale secured a $6-million grant from the Rockefeller Foundation to build a telescope featuring a mirror with a 200-inch diameter. After spending almost $1 million on unsuccessful attempts to form the mirror out of various materials including fused quartz, Hale decided to meet with optical experts at the Corning Glass Works in upstate New York to discuss the possibility of using a new glass blend called Pyrex. Pyrex offered an advantage: it expanded and contracted with changing temperature much less than ordinary glass did. The sharpness of the image formed by a telescopic mirror depends on the mirror having precisely the right shape. A Pyrex mirror would be less susceptible to the distortion and focusing problems that had plagued Hale's 100-inch telescope at Mount Wilson. Corning Glass accepted the challenge and, on its second attempt, succeeded in casting the mirror.

In 1936 the glass disc with a flat, rough surface was ready to be shipped across the country to Pasadena where technicians would grind and polish it. The disc was carried by train on a special flat car at speeds never exceeding 25 miles an hour. The project to build the world's largest telescope was easily the most famous scientific undertaking of the 1930s and captured the public's imagination. Schoolchildren of the era heard about it constantly. When the glass disc reached Indianapolis, 10,000 people were waiting along the railroad tracks to watch the glass disc go by. Guards were posted around the disc at night to prevent any damage.

After safely arriving at the Caltech optical workshops, the mirror was first ground to the approximate concave shape required. Then, using successively finer grit, the mirror was polished into a near perfect paraboloid shape. During the final stage of polishing, the grinders actually used their thumbs. The excruciatingly slow and painstaking process removed nearly 10,000 pounds of glass from the disc. Work on the mirror was interrupted in 1941 by World War II, and the mirror was put into storage for three years. Work resumed in September 1945, and the mirror was completed in 1947. On November 12, 1947, the 40-ton mirror was pushed from Pasadena to the top of Palomar Mountain by three diesel tractors. The 125-mile trip took 32 hours, and, once again, crowds gathered along the route to watch the mirror creep by. After the mirror was installed in the telescope structure, it took another two years to finish aligning and adjusting the mirror. In 1949, after 13 years of polishing, the telescope was, at long last, fully operational. It was officially named the Hale reflector in honor of George Hale, who did not live to see his dream telescope completed. In January 1949, none other than Edwin Hubble himself took the first photographic exposure.

The Hale reflector was the Hubble Space Telescope of its day and revolutionized astronomy, especially in the areas of stellar evolution and cosmology. When the telescope was first built, it was far from city lights, but as

The white dome holding the 200-inch Hale Reflector at the Palomar Mountain Observatory.

the cities of Los Angeles and San Diego have grown, the lights from the urban sprawl are endangering the usefulness of the telescope. Recent areas of research include searching for planets orbiting other stars, investigating large explosions in space, and hunting for asteroids that could threaten the Earth. In fact, nearly 10,000 asteroids have been discovered at Palomar Observatory including the large asteroid Icarus, the first discovery made with the telescope.

The huge dome that houses the telescope, one of the largest in the world, is as tall as a 12-story building. Painters use a clever technique to maintain the dome's brilliant white appearance: rather than move around the dome, they remain stationary, paint sprayer in hand, while the dome is rotated. The dome is divided into two sections. The lower concrete section holds computers, a library, storage rooms, various mechanical systems required for the operation of the telescope, and a vault containing old photographic plates taken with the telescope. The upper steel movable section weighs a thousand tons and rotates smoothly so that no vibrations are transmitted to the telescope. The most important requirement in the dome's design was its insulation so that changes in temperature inside the dome are minimized. To accomplish this, a four-foot gap separates the inner and outer concrete walls in the base of the building and the inner and outer metal walls of the moveable dome. During the day, warm air rises through the double walls and is replaced by cooler air. This circulating air keeps the dome cool in the daytime.

When you look at the telescope on a visit or in a photograph, it is easy to be confused as to what is the actual telescope and what is the supporting mount. The solid white cylindrical tubes are not the telescope tubes, but rather girders. The actual telescope is the open structure with all the triangular trusses situated between the girders. (A telescope does not have to be enclosed in a tube.) The mirror rests at the bottom of this open structure and the observer's cage is located at the top. The yoke—formed by the twin tubular girders that are joined together at one end by a cross member and at the other end by a horseshoe-shaped bearing—supports the weight of the telescope. Oil is pumped between the horseshoe and its supporting structure so that the horseshoe floats virtually friction free on a thin film of oil. This allows the horseshoe to rotate, which, in turn, enables the telescope to point in any direction, including straight up.

In addition to the 200-inch Hale reflector, Palomar Observatory is home to five other telescopes including a 60-inch, the 48-inch Samuel Oschin telescope (formerly known as the 48-inch Schmidt), a 24-inch, an 18-inch Schmidt telescope, and a small robotic telescope known as Sleuth and dedicated to the search for planets around other stars. Although these telescopes are not open to the public, the 48-inch Samuel Oschin deserves special mention because of its historical importance. This telescope saw first light in 1948 and, because of its wide field of view, was used to conduct the first Palomar Observatory Sky Survey, which pieced together a detailed map of the entire northern hemisphere night sky. It took ten years to complete this daunting task, but it was worth it. The survey became a standard reference for every major observatory around the world and allowed astronomers to pinpoint interesting celestial objects that merited further research. The survey even formed the basis for the Guide-Star Catalog used by the Hubble Space Telescope. The telescope performed a second digital survey from 1985 through 2000. Recently, the telescope discovered several small worlds (Quaoar, Orcus, and Sedna) in the Solar System's Kuiper Belt. These discoveries forced astronomers to finally agree on a definition of a planet in the fall of 2006.

Visiting Information

The observatory and visitor center are open to the public from 9 A.M. until 4:00 P.M. daily, except December 24 and 25, for free self-guided visits. As you enter the lobby of the visitor center, look back toward the entry to see two display cases. The case on the left holds a full-scale exact replica of Isaac Newton's first reflecting telescope. The case on the right shows a set of four small souvenir replicas of the 200-inch mirror, sold by Corning Glass, which could be used as coasters or ashtrays.

The museum consists of a single large room very similar to the museum at Mount Wilson, although marginally better. Photographs taken with the telescopes at Palomar Mountain encircle the room. Three small computer screens are available, one of which shows astronomical images captured at the observatory, another provides an update on solar system discoveries, and a third provides an interesting selection of short video clips and computer animations about the telescope. The best of these clips is a vintage 1948 movie called "The Big Eye Space Telescope." At another display, you can take a look at the atomic spectra of half a dozen chemical elements. A small gift shop is open on weekends. Plans are being developed to nearly double the size of the visitor center. The expansion will include a theater for popular lectures and movies, a larger space for exhibits, and a dome that will house a telescope for public use.

A short, easy hike up a paved walkway leads you to the imposing white dome, one of the largest in the world. In the entryway, you are greeted by a bust of George Hale, and the stairs to the left lead to a gallery where visitors view the telescope from behind a glass window. During the day, the telescope is pointed straight up so that the mirror rests on the bottom. This is done because glass is actually a very viscous fluid. If the mirror were stored at an angle, the glass would flow and, over time, the mirror's shape would be distorted. The viewing gallery has a 1/10-scale model of the mirror and diagrams explaining how the telescope works. On the back wall is a display case showing the wheels that rotate the dome.

By far the best way to see the giant telescope is to visit on a Saturday from April through October and buy a ticket to one of the excellent 90-minute guided public tours led by knowledgeable volunteers from the "Friends of Palomar" organization. The tours are offered at 11:30 A.M. and 1:30 P.M. and tickets are sold in the gift shop beginning at 10:00 A.M. on the day of the tour. The tickets are available on a first-come, first-served basis, and no reservations are accepted. Each tour is limited to 25 people and often sell out, so you better get there early to claim your ticket. Tickets are $5 for adults, $3 for seniors age 62 and older, and $2.50 for children ages 6 to 16. Families with two adults and up to four children can buy a family ticket for $12. The tours make a stop at the Testbed Interferometer, an instrument that works by combining the light from three small telescopes 110 meters apart. Then it's on to the dome. After stops in the entry for a chat about Hale and in the viewing gallery for a discussion about the scope, the guide unlocks the door to the main room of the dome where you will get an absolutely magnificent view of the legendary telescope. Photographs do not capture the enormous scale of the instrument. The guide uses a model to demonstrate how the telescope is maneuvered so that it can view any point in the sky. You will be led to the upper level for yet another spectacular view. If you're lucky (as I was on my visit), you might get to peek through the window at the control room where the astronomers actually work. The tour ends with a walk around the catwalk on the outside of the dome. From this vantage point, the guide points out the other telescopes on the mountain. A warning to visitors who, like me, have a fear of heights—the catwalk is a metal grate that you can see through. It's perfectly safe, but a little scary. Just don't look down! (According to the guide, a popular prank to play on graduate students new to the observatory is to lower one of the grates a few inches and lead them on a heart-stopping walk around the catwalk.) After the tour, walk around the outside of the dome where you will find the concrete disk that was used to mimic the weight of the mirror while the telescope's mounting was pieced together.

Depending on traffic, Mount Palomar is a two- or three-hour drive from Los Angeles and about a two-hour drive from San Diego. The

Website: www.palomar-observatory.org
Telephone: 760–742–2119 (public information recording)
          760–742–2111 (public affairs office)

winding mountain road is popular with motorcycle enthusiasts who enjoy the thrill of taking the perfectly banked curves at the maximum possible speed, so watch out for the bikers.

The National Solar Observatory, Sunspot, New Mexico

As the name suggests, the National Solar Observatory (NSO) specializes in observing our nearest stellar neighbor, the Sun. Here, basic research is done on such topics as the mechanisms governing small changes in the Sun's brightness and the properties and behavior of the Sun's magnetic field. The observatory keeps a watchful telescopic eye on the Sun, collecting data that is used to make a daily forecast of solar activity. Why should we care about what's happening on the Sun? Solar flares and prominences can disrupt radio communications, radar, satellite operation, and the electric power grid. The NSO telescopes can give us advance warning of these events.

The public can view up close two telescopes at the NSO: the Dunn Solar Telescope and the Evans Solar Facility. Standing 13 stories tall, the Dunn Solar Telescope is the dominant structure on the mountaintop. As is the case with other solar telescopes, the Dunn has a tower design to get the optics above the distortions caused by heat released by the ground. Only about a third of the telescope is visible above ground with the rest of the structure extending more than 200 hundred feet below the ground. The entire telescope is longer than a football field. At the top of the tower two mirrors reflect sunlight down the main tube where it hits the 64-inch primary mirror. This mirror focuses the light and reflects it back up to ground level where a final set of mirrors direct the light out of the tube and into the experimental stations. The Dunn's main telescope tube, weighing more than 200 tons, is suspended from the ceiling from a ring-shaped container holding ten tons of mercury. The mercury has very little friction, which makes it easy to rotate the heavy tube. Scientists use this telescope to conduct research on sunspots, flares, magnetic fields, and granulation.

The Evans Solar Facility houses a special 16-inch telescope called a coronagraph, which is used to study the corona, the outermost layer of the Sun's atmosphere. To understand what a coronagraph does, think of a total solar eclipse. During the eclipse, the Moon moves in front of the disk of the Sun and blocks the light, making the faint corona visible. A coronagraph simulates an eclipse by using a disk built in to the optics of the telescope. The facility also houses a 12-inch coelostat, an instrument consisting of two flat mirrors turned by a motor to follow the Sun across the sky. In addition to observing the corona, astronomers use this facility to study solar flares, filaments, and the external solar magnetic field.

Visitors may also view the Hilltop Dome from the outside. This dome contains several telescopes that do nothing but watch the Sun all day, taking pictures at one-minute intervals. The telescopes take two sets of photographs: one set in the visible part of the spectrum just as the human eye would see it, and another set at a particular wavelength produced by hydrogen atoms in the Sun. This second method produces images that bring out much more detail than normal images do. Together, these images form a daily record of the Sun's behavior and are used to discover when something unusual is happening so that astronomers can take a closer look.

Finally, the first telescope dome built on this mountaintop in 1950 was the Grain Bin Dome, so-named because the dome is actually a grain bin ordered right out of the Sears catalog. From 1951 through 1963, the telescopes housed here were used to watch the edge of the Sun for solar flares. Today, it holds a small telescope for night viewing.

Visiting Information

The grounds are open daily, free of charge, from dawn to dusk. The Visitor Center contains a small, but good, museum with interactive demonstrations and exhibits on interference, refraction, diffraction, polarization, atomic spectra, and telescopes. Don't miss looking at yourself through the infrared camera, and be sure to examine the live pictures of the Sun on TV monitors. In front of the Visitor Center is a five-foot diameter armillary sphere and sundial accompanied by a plaque explaining how to read it. A brochure for a self-guided walking tour is available in a box outside the Visitor Center or inside at the reception desk. The walking tour takes you on a short quarter-mile loop around the facilities and includes stops at the Evans Solar Facility, the Dunn Solar Telescope, a scenic view over the Tularosa Basin, the Hilltop Dome, and the Grain Bin Dome. The telescopes are dimly lit during the day, and the view is somewhat disappointing. During the summer, the Visitor Center is open daily from 10:00 A.M. to 6:00 P.M. There is a $3 charge for the museum; $7 for a family. From Memorial Day through Labor Day, there are free guided tours daily at 2:00 P.M. In the spring and fall, the tours are on weekends only. In winter, its best to call ahead to make sure the observatory is open and accessible. About a mile and a half down the road from the NSO is Apache Point Observatory. None of the buildings here are open to the public, although visitors are welcome to stroll the grounds from 7:00 A.M. until 5:00 P.M. A brochure describing the telescopes at Apache Point is available at the visitor's center at the NSO.

The NSO is located in southeastern New Mexico, about a one-hour drive from Alamo-

Website: www.nso.edu
Telephone: 505–434–7190

gordo. The road that takes you to the entrance, NM 6563, is the wavelength, in a unit called Angstroms, of what astronomers call the Hydrogen-alpha, the wavelength that hydrogen gas emits or absorbs when at a temperature of 18,000°F. Isn't that clever? Also, Sunspot is just the cute name for the group of buildings that form the NSO. It's not a real town.

Kitt Peak National Observatory, Tucson, Arizona

High on a mountaintop 7,000 feet above Arizona's Sonora Desert, Kitt Peak National Observatory is home to twenty-five optical and two radio telescopes, the largest collection of telescopes in the world. At the most visited observatory in the world, the spectacular natural setting is part of the attraction. Kitt Peak is named in honor of Philippa Kitt, the sister of George J. Roskruge, a Pima County surveyor who named many places in southern Arizona. To keep light pollution to a minimum, the city of Tucson and the surrounding county have passed ordinances requiring reflectors on area lighting. A camera at Kitt Peak occasionally scans the skyline toward Tucson looking for bothersome bright spots, which reflectors then subdue. The National Optical Astronomy Observatory (NOAO), funded by the National Science Foundation, oversees site operations on Kitt Peak.

Because Kitt Peak is located within the Tohono O'odham Indian Reservation the tribe's permission was required before any telescopes could be built on the mountain. The tribal leaders were reluctant at first because Kitt Peak is a sacred mountain and they didn't want to have the peace and quiet of their deities disturbed. They mistakenly thought that the astronomers might be shooting rockets off the top of the mountain, but their fears were assuaged by arranging for a group of tribal elders to look at the Moon and the stars through the 36-inch telescope at the Seward Observatory in Tucson. This, the elders were told, is what the astronomers had in mind for Kitt Peak. Dubbing the astronomers as the "people with the long eyes," the elders gave their consent.

A perpetual lease with annual payments of $25,000 was drawn up and signed by the National Science Foundation and the Tribal Council. The lease runs "as long as the property is used for astronomical study and research and related scientific purposes" and features a couple of interesting provisions. First, the caves near the summit may not be entered because the Papago god EE-EE-Toy might be relaxing inside. Second, the Visitor Center is required to sell Indian crafts without taking a cut of the profits. With no middleman, the prices are lower, and visitors can find some good bargains on Native American jewelry and basketry.

Kitt Peak National Observatory. The dome of the 4-meter telescope is on the right.

Three telescopes are publicly accessible at Kitt Peak: The Mayall 4-meter (13-foot) telescope, the McMath-Pierce Solar Telescope, and the 2.1 meter (about 6.5 foot) telescope. The telescopic skyline of Kitt Peak is dominated by the eighteen-story-tall dome of the Mayall 4-meter telescope. This largest telescope at Kitt Peak is among the largest in the world. The telescope, completed in 1973 at a cost of $10 million, is named in honor of Nicholas U. Mayall, the director of Kitt Peak from 1960 to 1971. The primary mirror, weighing 15 tons, is made of fused quartz, a material that resists thermal expansion. The mirror is polished to an accuracy of a millionth of an inch and is covered during the day to protect it from dust and debris. The 92-foot-long telescope is cradled in a blue horseshoe-shaped equatorial mount securely anchored to a concrete pier that is completely separate from the building and the dome. Thus, the telescope will not feel any image-blurring vibrations when the massive 500-ton dome is moved. During the day, the dome is cooled to the temperature predicted for the upcoming night. This prevents image distortion from ripples of heat coming off the floor. The Mayall has been used to study the structure, rotation, and the distribution of galaxies, to observe supernovae in distant galaxies, and to search for brown dwarf stars in our own galaxy. The telescope has two visitor galleries, one to view the telescope, and another for scenic panoramas of the surrounding territory.

One of the most unusual telescopes to be found anywhere and certainly the telescope that is most recognizable in photographs of Kitt Peak is the triangular McMath-Pierce Solar Telescope; the largest solar telescope in the world, it is named in honor of Dr. Robert McMath, who pushed to get the telescope built, and Dr. Keith Pierce, who made significant contributions to our understanding of the Sun. Atop the 11-story tall vertical tower, three primary mirrors, called heliostats, track the Sun across the desert sky. The largest mirror, nearly seven feet across, is flanked by two secondary mirrors measuring three feet in diameter. This trio of mirrors, actually three telescopes in one, enables the telescope to be used for three independent research projects at the same time. These mirrors reflect sunlight down the 500-foot-long slanted section of the telescope; 200 feet of the section is visible above ground while the other 300 feet extends underground. At the bottom of the tunnel, the light hits a concave mirror that forms the image. The light is then reflected about two-thirds of the way back up the tunnel to one of two flat mirrors mounted diagonally on rails. These mirrors can be moved so that the light can be reflected to different instrument rooms. This arrangement produces a sharp image of the Sun almost three feet across or a spectrum 70 feet long. Auxiliary equipment allows observations of the solar corona and flares.

The McMath-Pierce Solar telescope was specifically designed to counteract the problem of heating, the bane of solar astronomers. Anybody who has ever burned holes in paper with a lens understands that a focused beam of sunlight produces a significant amount of heat. Heating air inside a telescope by an intense beam of sunlight causes convection currents, which move blobs of air, which, in turn, create turbulence that can distort the images. This effect is similar to the rippling air directly above hot pavement that can be seen on a hot summer's day. To remedy this problem, just inside the visible white walls of the telescope is a skeleton of 25,000 feet of copper tubing circulating tens of thousands of gallons of a chilled antifreeze-water mixture. The tubing is bonded to the 14,000 copper panels that form the outer skin of the telescope. This keeps the air inside the slanting shaft ten degrees cooler than the air outside. The air inside is motionless, and the Sun's image is undisturbed.

The final telescope that is open to visitors is the 2.1-meter telescope, the first large telescope built at Kitt Peak. The Pyrex mirror weighs 3,000 pounds and is polished to an accuracy of four millionths of an inch. The telescope has a short, stubby appearance compared to other telescopes. This arrangement gives the telescope a wider field of view, which allows it to see a larger piece of the sky at one time. The design is also economical because it requires less material and smaller domes. Several significant discoveries have been made with this telescope, including the first gravitational lens, the first clumps of gas between the galaxies, and the first pulsating white dwarf. The telescope was also used to conduct the first comprehensive study on how often sunlike stars are part of binary star systems. Today, the telescope is used to make observations in the near-infrared part of the spectrum. This gives astronomers the ability to study the distribution of dust in the universe.

Visiting Information

After completing your journey to the top of the mountain, stop first at the Visitor Center where you can get oriented and ponder the exhibits on astronomy and telescopes. You will also want to grab a map for a self-guided tour or join up with a guided tour. The gift shop's handy booklet, *A Visitors Guide to the Kitt Peak Observatories*, describes each telescope along with other points of interest and scenic vistas. There is no food service on the mountain, so you may want to bring a picnic basket and eat lunch outside at the Visitor Center or at the picnic grounds next to the Very Long Baseline Array (VLBA) telescope.

The southern wall of the Visitors Center is decorated with a beautiful tile mosaic representing the ancient Mayan El Caracol (the snail) observatory located at Chichén Itzá on Mexico's Yucatan peninsula. The observatory is surrounded by a sky holding a stylized Sun, Moon, and planets. At the upper left hand corner of the mosaic is a replica of a Mayan *glyph,* a basic unit in Mayan writing consisting of a pictogram enclosed by a square frame. Listen to the nearby audio recording for a detailed explanation of the symbolism. The Visitor Center plaza has a sundial inscribed with the exact latitude and longitude of Kitt Peak's summit. In the parking lot is a giant donut-shaped disk of concrete with the same dimensions and weight as the actual mirror of the 4-meter Mayall telescope. This disk was used during the assembly of the telescope as a "stand-in" for the real mirror. Just to the southwest of the Visitor Center is a shaded gardenlike area where a plaque mounted on a boulder honors the Tohono O'odham culture. The plaque depicts a man shouldering the tribe's "man in the maze" symbol. The man is Elder Brother Iitoi finding his way home from the top of Baboquivari Peak (this thumblike peak is clearly visible from Kitt Peak), the center of the universe to the Tohono O'odham tribe.

The Visitor Center is open from 9:00 A.M. until 3:45 P.M. daily except Thanksgiving, Christmas, and New Year's Day. There is a suggested donation of $2.00 per person. Three guided tours originating from the Visitor Center are offered daily. A 10:00 A.M. one-hour tour goes to the solar telescope, an 11:30 one-hour tour to the 2.1 meter telescope, and a 1:30 P.M. 90-minute tour goes to the 4-meter telescope. Cost for the guided tours is $2.00 for adults, $1.00 for children 6 to 12, and children under six are free.

Kitt Peak offers two nightly observing programs: a general three-hour observing program and a more advanced all-night observing program. The general observing program includes a light meal, an introduction to the night sky, and viewing objects through binoculars and a 16-inch reflecting telescope. The advanced program is geared more toward serious amateur astronomers who are interested in using large telescopes and state-of-the-art equipment. Both of these popular programs are by reservation only and sell out nearly every night. To secure a spot, call the number below at least three weeks in advance for the general program and at least a month in advance for the all-night program. The cost for the general program is $39 for adults and $34 for seniors and students. A $10 deposit is required to hold your spot. The cost for the advanced program is $375 per night for two people. These programs are offered nightly except for July 15 to August 31, which is the rainy

> Website: www.noao.edu
> Telephone: 520–318–8200 (recorded information)
>   520–318–8726 (reservations)

season. Kitt Peak is about 60 miles southwest of Tucson, Arizona, and is approximately a 90-minute drive.

Keck Observatory, Mauna Kea, Hawaii

The two largest optical telescopes in the world as of this writing, Keck I and Keck II, sit atop Mauna Kea (Hawaiian for "white mountain"), a massive dormant volcano on the island of Hawaii. Mauna Kea's elevation and location make it one of the very best astronomical observing sites on the surface of the Earth. At an altitude of nearly 14,000 feet, the summit of Mauna Kea sits above 40 percent of the Earth's atmosphere allowing astronomers to enjoy an average of 240 clear, cloudless nights a year for observing. (If measured

Jeanne Hatch/Shutterstock

The Mauna Kea Observatories. The twin domes of the Keck Telescopes are visible at the right.

from its base, 20,000 feet below the surface of the Pacific Ocean, Mauna Kea is the tallest mountain on Earth.) The gentle slopes of the volcano make the summit easily accessible; there aren't many mountains this tall you can drive up. The elevation also puts the peak above 90 percent of the water vapor in the Earth's atmosphere. Water vapor absorbs infrared radiation coming to the Earth from outer space and makes infrared observations from low elevations impossible. But at Mauna Kea, little infrared has been absorbed, and the telescopes can detect it. Several telescopes on Mauna Kea are specially designed for infrared observations. Then, there is the location. At a northern latitude of only 20 degrees, most of the northern and southern skies are visible. No nearby mountain ranges can produce dust or roil the atmosphere, and no nearby large cities can produce light pollution. Plus, the volcano is surrounded by an ocean that stabilizes the temperature. The extremely dark night skies and thin, stable air of Mauna Kea make it a perfect place for astronomy.

The twin Keck telescopes each have a primary mirror ten meters (33 feet) in diameter, twice the width of the Hale reflector at Palomar Mountain with four times the light-gathering ability. Unlike the Hale reflector, the Keck mirrors are not made from a single piece of glass; rather, 36 individual hexagonal pieces fit together in a honeycomb pattern. The polished surface of each segment is so smooth that if the segments were scaled up in size to the diameter of the Earth, the imperfections would only be three feet high. Sensors at the edge of each segment measure its exact orientation every half second. The data is sent to a computer which calculates any necessary adjustments due to gravity, the wind, or vibrations from the motion of the telescope. The computer then sends instructions to a system of actuators which perform the adjustments. In this way, the segments cooperate to form a single concave paraboloid surface accurate to within a millionth of an inch. An adaptive optics system has been installed on the telescopes that cancels out atmospheric distortions. This has improved the resolution of the images by a factor of ten.

The Keck telescopes rest on an altitude-azimuth or "altazimuth" mount, which enables the instruments to pivot about a vertical and horizontal axis. In designing the mount, computer analysis was used to maximize the strength and rigidity of the structure by using the minimum amount of steel. This not only saves money, but it also reduces the deformations in the structure as a result of its own weight. The mounting and optical design of the telescope allow it to be housed in a dome that is relatively small. In fact, the dome for the ten-meter Keck is significantly smaller than the dome for the five-meter Hale Reflector. Giant air-conditioners run continuously during the day to keep the interior of the dome at night-time temperatures, usually at or below freezing. Keeping the temperature constant minimizes the expansion and contraction of the steel and glass.

Keck I saw first light in 1993 and Keck II in 1996. The telescopes, virtually identical, are housed in the same observatory facility but in separate domes. Having two telescopes not only doubles the observing time available to astronomers but also allows the twin instruments to be used together as an interferometer. In this way, the telescopes can achieve a resolving power equivalent to a single mirror 85 meters wide, the distance between the telescopes. The Keck-Keck Interferometer is part of NASA's "Origins" program— an effort to understand how planetary systems form. The interferometer will detect planet-forming dust around stars, obtain images of protoplanetary disks, and discover planets orbiting sunlike stars.

The Keck Telescopes are owned and operated by the California Association for Research in Astronomy, a partnership that includes Caltech, the University of California, and NASA. The telescopes are named in honor of William Myron Keck, founder of the Superior Oil Company. The W. M. Keck

The 10-meter Keck Telescope, the largest optical telescope in the world.

Foundation provided grants totaling more than $140 million to construct the telescopes.

The Keck telescopes are not alone on top of Mauna Kea. Indeed, the summit has become an international center for astronomy with 13 telescopes operated by countries from around the world. However, the only other telescope that is equipped with a visitor gallery is the University of Hawaii's 2.2-meter (about 7-foot) telescope. The first big telescope constructed on Mauna Kea, it saw first light in 1970, and its successes demonstrated to the world Mauna Kea's excellent observing conditions. The most significant discovery made with this telescope was the existence of what became known as the Kuiper Belt. Before 1992, the only object known to exist beyond the orbit of Neptune, aside from comets, was Pluto. The first Kuiper Belt object was discovered with this telescope in 1992, and we now know that the outer solar system is teeming with tens of thousands of small bodies. Pluto is now regarded as the largest Kuiper Belt object.

Another telescope that visitors can tour by prior arrangement is the 8.2-meter (27-foot) Subaru Telescope. *Subaru* is the Japanese name for the visible star cluster known as the Pleiades, a group of stars the Japanese have admired for more than a thousand years. The word *Subaru* means "to get together," a good name for an international research facility. Also visible on the summit are the eight 6-meter dishes of the Submillimeter Array and the 25-meter radio dish of the Very Long Baseline Array is located three miles below the summit. Other telescopes whose domes can be seen on Mauna Kea include the NASA Infrared Telescope Facility, the United Kingdom Infrared Telescope, the Gemini Northern Telescope, and the James Clerk Maxwell Telescope.

Visiting Information

The Mauna Kea Observatory is publicly accessible, but it is not easy. You can drive to the top of Mauna Kea on your own but the road is unpaved, steep, rough, and winding. Only four-wheel-drive vehicles are permitted on the road above the Visitor Information Station, and few, if any, rental car companies allow you to drive their rentals on this road. Extreme caution must be used when driving on the summit road, especially on the descent where you should be on the watch out for slides and loose gravel. Use low gear and keep your speed under 25 miles per hour. Avoid driving at sunrise and sunset because you have to drive directly into the Sun on the switchbacks which may blind you to oncoming traffic. Your safest and best option is to sign up for a commercial tour that takes you to the summit. A list of authorized tour operators is available on the website below. Keep in mind that during the winter months, storms can dump several feet of snow on the summit making the road impassable. It is a good idea to call ahead to check the weather conditions. If you decide to make the trek to the top, bring warm clothes (temperatures at the summit can fall to 20°F during the day), sunscreen, water, and sturdy shoes.

Mauna Kea is located on the "Big Island" of Hawaii. To get there, get on Route 200, also known as Saddle Road, a narrow road that winds its way between Mauna Loa and Mauna Kea. At Mile 28, turn onto the paved road that will take you up to the Visitor Station at the Onizuka Center for International Astronomy. This is where astronomers and technicians acclimate themselves to the altitude. The Visitors Center is 34 miles from Hilo, with a driving time of between 60 and 90 minutes. A stop of at least 30 minutes at the Visitor Station is recommended to allow your body to get used to the altitude.

The Visitor Station offers videos to watch, brochures and handouts to read, and displays to explore. The center also has several good quality telescopes for visitors to look through. During the day, one telescope is equipped so that the Sun and the solar spectrum can be viewed. Stargazing programs are offered nightly from 6:00 to 10:00 P.M. The Visitor Station offers free guided tours of the summit on weekends at 1:00 P.M., but you must provide your own four-wheel-drive transportation.

At the summit, the Keck Observatory has a visitor's gallery where you can watch a 15-minute video on the telescopes and get a partial view of the Keck I telescope and dome. The Keck gallery is usually open from 10:00 A.M. to 4:00 P.M. on weekdays only. The University of Hawaii's 2.2-meter telescope is about half a mile down the road. Its visitor gallery is usually open from 9:30 A.M. to 3:30 P.M. from Monday through Thursday.

To tour the Subaru Telescope, reservations must be made at least a week in advance. These half-hour tours include a look at the telescope and the facility used to recoat the primary mirror. You may sign up for a tour on the Subaru telescope website. The tour schedule is posted three months in advance, reservations are first-come, first-served, and space on the tours is limited, so reserve your spot as soon as possible. Some of the commercial tour operators may include this telescope in their package.

You may want to couple your trip to Mauna Kea with a visit to the Imiloa Astronomy Center of Hawaii. This new $28 million facility includes

Websites: www.keckobservatory.org
www.naoj.org (This website gives you information on the
Subaru Telescope tours.)
www.ifa.hawaii.edu/mko (This website has information on
visiting Mauna Kea including a list of commercial tour
operators and has a link to the Mauna Kea weather center
that will allow you to check the weather forecast.)

Telephone: 808–961–2180 (This is the Visitor Information Station number.)

interactive exhibits, a planetarium, an astronomy store, and a café. The center is located in Hilo on the campus of the University of Hawaii at Hilo and is open Tuesday through Sunday from 9:00 A.M. until 4:00 P.M. Admission is $14.50 for adults and $7.50 for children 4 to 12.

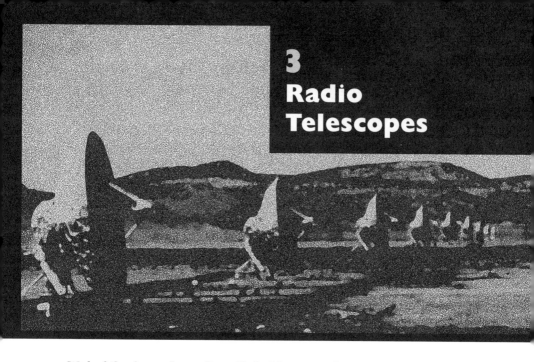

3
Radio Telescopes

We had the sky, up there, all speckled with stars, and we used to lay on our backs and look up at them, and discuss about whether they was made, or only just happened. From *Huckleberry Finn* by Mark Twain

Aside from visible light, the other realm of the electromagnetic spectrum that can penetrate through the Earth's atmosphere is radio. Radio waves can be longer than a football field or as short as a football. Because an electronic radio emits sound, some people have the mistaken idea that radio waves are sound waves and that radio telescopes are like big ears listening to signals from space. Regrettably, this misconception was reinforced when, in the movie *Contact*, Jodie Foster donned earphones to listen for an extraterrestrial message. An electronic radio converts the radio waves into sound waves, but the radio waves themselves are electromagnetic waves.

Of course, the human eye cannot see radio waves so astronomers have to use instruments called radio telescopes that can "see" radio waves. Radio telescopes are large parabolic dishes that collect radio waves. Radio telescopes have to be big because the energy of radio waves is very small. According to the late astronomer Carl Sagan, "The total amount of energy from outside the solar system ever received by all the radio telescopes on the planet Earth is less than the energy of a single snowflake striking the ground." Sagan said that in 1980, so maybe by now we're up to a snowflake's worth of energy.

Although radio telescopes look a lot different than optical telescopes, their function is basically the same: to collect and focus electromagnetic radiation. The big dish collects and reflects the radio waves just like the mirror of a reflecting telescope. The dish's parabolic shape forces the waves to focus at a point. An antenna located at this point converts the radio waves into a weak electric signal. This signal is amplified and then fed to a computer for analysis. By scanning the telescope across a radio source in the sky, variations in the frequencies can be observed. This can provide information about the structure, composition, and motion of the source. The detection of radio waves coming from the depths of space has led to a host of astronomical discoveries, including objects like quasars that were previously unknown.

Most radio telescopes are "steerable" in the sense that they can be tilted and turned to any part of the sky. However, there is an upper limit to the size of a steerable radio telescope for the same reason there was an upper limit to the size of a lens: as an object is scaled up in size, its weight increases faster than its strength. The largest fully steerable radio telescope in the world is the Robert C. Byrd Radio Telescope at Greenbank, West Virginia. With a diameter of 330 feet, the Byrd Radio Telescope is about as big as a steerable radio telescope can get. The largest single-dish radio telescope in the world is the 1,000-foot Arecibo telescope in Puerto Rico, but this telescope is set into the ground and cannot be steered. Like optical telescopes, most radio telescopes are painted white so that much of the sunlight is reflected. This minimizes the heating of the dish, which, in turn, minimizes the expansion of the metal.

The amount of detail a radio telescope can see—a property called the resolution or resolving power—depends on how big it is. A large radio telescope has about the same resolving power as a human eye. To improve the resolution, radio astronomers use an entire group or "array" of dishes and combine their signals. A group of telescopes combined in this way is called an interferometer, and the result is a resolving power equivalent to that of a single telescope with a diameter equal to distance across the array. In this way, many smaller telescopes can act as one giant dish.

Optical telescopes are usually located far from population centers so as to avoid light pollution. Similarly, radio telescopes are located in isolated areas to avoid electromagnetic interference from radio, TV, and radar. In contrast to optical telescopes that are perched on mountaintops, radio telescopes are often placed in valleys so that the surrounding mountains can shield against interference. Radio astronomy has one advantage over optical astronomy: clouds and rain do not affect the observations. Also, sunlight does not overwhelm radio waves coming from space so radio astronomy can be done night and day.

The first radio antenna that detected an astronomical radio source was built by Karl Guthe Jansky in the early 1930s. Jansky, an engineer who worked for Bell Telephone Laboratories in New Jersey, was given the job of finding sources of static that might interfere with radio telephone service. To detect these troublesome sources, he built a 110-foot wide, 20-foot tall radio antenna mounted on a turntable that rotated on a set of four Ford Model-T tires. The antenna became known affectionately as "Jansky's merry-go-round." The signals were recorded using a pen-and-paper recording system housed in a nearby shed. After several months of observing, Jansky recognized that much of the static originated in thunderstorms, but there was a faint steady hiss of static from a source he could not readily identify. After studying some astronomical maps, he realized that the direction of the source coincided with the direction of the center of the Milky Way Galaxy. Jansky's mysterious source of static was the galaxy itself.

The first true radio telescope was built by Grote Reber, a ham radio operator who had studied radio engineering and worked for various radio manufacturers in the Chicago area. When Reber learned of Jansky's discovery of radio waves from the Milky Way Galaxy, he wanted to do some follow up research to learn more about these mysterious cosmic radio waves. During the 1930s, Reber tried to land a job at Bell Labs and at astronomical observatories to pursue this research, but during the Great Depression nobody was hiring. Reber decided to do it himself. In 1937, he put together a parabolic dish that measured nine meters across and used it to conduct the first sky survey at radio wavelengths. His original dish can be seen at the Green Bank Observatory in West Virginia. The new field of radio astronomy really took off after World War II with major improvements in technology. Below, I describe in chronological order three of the major radio astronomy observatories in the United States and Puerto Rico.

The National Radio Astronomy Observatory, Green Bank, West Virginia

The National Radio Astronomy Observatory (NRAO) was founded in 1956, and the first of its radio telescopes was built at the Green Bank site. Today, the NRAO designs, builds, and operates large radio telescopes here and at several sites around the world, including a 39-foot radio telescope at Kitt Peak, the Atacama Large Millimeter Array (ALMA) near Santiago, Chile, and New Mexico's Very Large Array. At the Green Bank site, extreme measures have been taken to shield the telescopes from interference from earthly radio sources. For example, the Green Bank Observatory sits on a 13,000-square-mile plot of land that has been set aside by the Federal Communications Commission as a "Radio Quiet Zone." Radio transmitters inside the zone must transmit their signals in a direction away from the telescopes, often at reduced power. Even pine trees with needles of just the right length needed to block electromagnetic interference at the wavelengths used by the telescopes have been planted around the surrounding area. Of course, it is difficult to always anticipate possible sources of interference. At one time, the telescopes were picking up interference from transmitters attached to flying squirrels that had been tagged by the U.S. Fish and Wildlife Service.

The NRAO has a cast of several radio telescopes, but the star of the show is the 330-foot Robert C. Byrd Green Bank Telescope (GBT), named in honor of the white-haired senator from West Virginia who helped secure funding. Standing 43 stories tall and featuring a dish with a collecting area of 2.3 acres, the GBT is the largest fully steerable radio telescope in the world. Although most conventional radio telescopes have a series of supports in the middle of the dish, the GBT is unique in that the dish is completely unblocked. This innovative design increases the useful area of the dish and improves its performance. Each of the 2,000 panels that make up the surface can be adjusted so that the shape of the dish is accurate to a hundred-millionths of a meter.

Other radio telescopes of note at the NRAO include an interferometer array of three 85-foot dishes used for research on pulsars, a 140-foot telescope used to study the Earth's ionosphere, a 40-foot telescope used for educational purposes including student research, and a horn antenna used to observe Cygnus X-1, a suspected black hole. In addition, several radio telescopes at the NRAO are of high historical interest: first, a full-scale replica of Karl Jansky's original radio antenna built at Bell Laboratory; second, Grote Reber's first radio telescope, which Reber donated to the observatory; and third, the horn antenna that was used to discover the 21-cm. (8.3-inch) wavelength radio waves emitted by hydrogen—an extremely important wavelength used by radio astronomers.

David Kay/Shutterstock

The 330-foot Robert C. Byrd Radio Telescope at Green Bank, West Virginia, the largest fully steerable radio telescope in the world.

Another bit of Green Bank history: the first attempt to detect signals from an extraterrestrial civilization, nicknamed Project Ozma after the princess of Oz, was undertaken here by astronomer Frank Drake in 1960. Using the 140-foot telescope, Drake listened for signals from the nearby stars Epsilon Eridani and Tau Ceti for several weeks, but, alas, no signals from ET.

Visiting Information

The NRAO is visitor friendly and has a plethora of activities from which to choose. From Memorial Day through Labor Day, the NRAO is open to the public from 8:30 A.M. to 7:00 P.M. daily and offers free public tours on the hour from 9:00 A.M. through 6:00 P.M. From Labor Day through October, the hours are the same, but the facility is closed on Mondays and Tuesdays. For the rest of the year, the NRAO is open Wednesday through Sunday from 10:00 A.M. to 5:00 A.M., with guided tours at 11:00A.M., 1:00 P.M., and 3:00 P.M. The NRAO is closed on major holidays. The guided tours include demonstrations, a film on radio astronomy, and a bus tour of the telescopes. Self-guided walking or bicycling tours may be taken anytime. Special events including Star Parties, Star Lab Thursdays, and Film Fest Fridays are scheduled throughout the year. Check the website for schedules and details. Visitors interested in a behind-the-scenes look at the NRAO may want to check out the High Tech Wednesday tours.

You can easily spend an hour or two learning about radio astronomy from a variety of hands-on exhibits and displays at the Green Bank Science Center. The Galaxy Gift Shop stocks NRAO souvenirs, and you can grab a bite to eat at the Starlight Café.

The NRAO is located on Route 92/28 about 25 miles north of Marlington, 53

> Website: www.gb.nrao.edu
> Telephone: 304–456–2150

miles south of Elkins, and 65 miles west of Staunton, Virginia. Specific directions from various nearby cities can be found on the website.

Arecibo Radio Telescope, Arecibo, Puerto Rico

The Arecibo Observatory (also known as the National Radio and Ionosphere Center) is home to the largest single dish radio telescope in the world. The 1,000-foot-diameter dish is nestled in a natural bowl-shaped valley called a *karst*. The spherical reflecting surface, made up of nearly 40,000 perforated aluminum panels, each measuring approximately six feet long by three feet wide, is supported by a mesh of steel cables covering a total area of about 20 acres. The surface collects radio waves from space and focuses them onto an antenna held by the triangular platform suspended 450 feet above the reflector. The platform is supported by 18 cables strung from three 250- to 300-foot towers firmly anchored to the surrounded hilltops.

The Arecibo Observatory was the brainchild of Professor William E. Gordan of Cornell University. Gordan was interested in building a radio telescope to study the ionosphere, a layer of the Earth's upper atmosphere where atoms and molecules are ionized by solar radiation. The ionosphere is of practical importance mainly because it effects the propagation of radio

The 1,000-foot Arecibo Radio Telescope in Puerto Rico, the largest radio telescope in the world.

waves around the Earth. Gordan agreed to changes in his original vision so that the telescope would be a more versatile astronomical research instrument. Construction began in 1960, and the observatory officially opened on November 1, 1963.

A number of highly significant scientific discoveries have been made by scientists working with the Arecibo telescope. One of the earliest discoveries took place in 1965 when the telescope was used to establish the time it takes the planet Mercury to spin on its axis once. The Mercury "day" was measured to be 59 days rather than the 88 days that had been previously deduced. In 1974, astronomers Russell Hulse and Joseph Taylor discovered the first pulsar in a binary star system. This discovery, in turn, yielded indirect proof of the existence of gravity waves, a prediction made by Einstein's General Theory of Relativity. Hulse and Taylor were awarded the 1993 Nobel Prize in Physics for their discovery. In the early 1990s, Polish astronomer Aleksander Wolszczan used the Arecibo dish to discover the first planets outside our own solar system.

Arecibo has also played an important role in the (so far) unsuccessful search for extraterrestrial intelligence (SETI). The Arecibo radio telescope has, on several occasions, been used to search for radio signals from an advanced civilization. In 1992, the U.S. government, under the auspices of NASA, funded a project called the microwave observing program (MOP) which searched 800 nearby stars for intelligent signals. Congress unfortunately criticized the program as a waste of taxpayer's money and cancelled it after only a year. Project Phoenix, funded with private money, arose from the ashes of the MOP Project and continued the search. Officials chose Dr. Jill Tarter, who had worked on the MOP project, to lead Project Phoenix; she currently directs the SETI Institute. The Ellie Arroway character in *Contact* is loosely based on Tarter.

In 1974, at a ceremony to commemorate the completion of a major upgrade, the telescope, rather than passively listening for a message, beamed a message of its own out into the depths of space. The message, consisting of 1679 binary pixels of information, included the numbers one through ten, information about the DNA molecule, and simple graphic representations of a man, the DNA double helix, the solar system, and the Arecibo radio telescope. The message was aimed toward a cluster of stars 25,000 light-years away. In the extremely unlikely event that an intelligent being receives and responds to the message, we shouldn't expect a reply for another 50,000 years. Like stuffing a message into a bottle and tossing it into the ocean, the message was more symbolic than a serious attempt to talk to E.T.

The Arecibo telescope has made frequent appearances in the movies and on TV. Most famously, the opening scenes of the movie *Contact* starring Jodie Foster were filmed at Arecibo, as were the climactic scenes in the James Bond movie *Golden Eye*. The observatory was also featured in the science fiction movies *Species* and *Survivor*. In the *X-Files* episode "Little Green Men," the lead character, Fox Mulder, is sent to Arecibo to see if contact had been made with extraterrestrials.

Visiting Information

The Arecibo Observatory is about a 90-minute drive from the San Juan airport. See the website for directions. The observatory has a visitor center with exhibits focusing on the theme "More Than Meets the Eye—Exploring the Invisible Universe." There are four main galleries at the visitor center: The Earth and Our Solar System Gallery and the Stars and Galaxies Gallery describe the objects that are studied with the telescope; the Tools and Technology Gallery explains how the telescope works; and The Work at Arecibo Gallery tells about some discoveries that have been made with the telescope and describes current areas of research. A 20-minute audiovisual presentation, "A Day in the Life of the Arecibo Observatory," tells the story of the people, from the cooks to the telescope operators, who make Arecibo happen. After the exhibits and the show, wander out onto the observing platform for a spectacular view of the giant radio telescope. The visitor center is open Wednesday through Friday from noon until 4:00 P.M.

> Website: www.naic.edu
> Telephone: 787–878–2612

and from 9:00 A.M. until 4:00 P.M. on weekends and holidays. Admission is $5.00 for adults and $3.00 for children and seniors. A gift shop and refreshment area is located in the visitor center.

Very Large Array, Socorro, New Mexico

Rising like giant white flower blossoms from the barren desert floor are the twenty-seven radio antennas of the aptly named Very Large Array (VLA). Fans of the movie *Contact* will recognize the VLA as the place where Ellie Arroway, played by Jodie Foster, first makes contact with an alien civilization. A number of scenes in the movie were in fact filmed at the VLA, although the canyon seen near the VLA in the movie is actually Canyon de Chelly in Arizona. The irony is that SETI (Search for Extraterrestrial Intelligence) research has never been done at the VLA because large, single dish radio telescopes are actually better suited for that purpose. Besides, the VLA is in such demand by astronomers for traditional research that no telescope time is available for SETI.

The VLA was completed in 1980 at a cost of nearly $80 million. The dishes can be moved using a specially designed transporter that lifts the 82-foot diameter, 230-ton telescopes from a pedestal, carries them to a new location along railroad tracks, and gently sets them down on a new pedestal with a precision of a quarter of an inch. Each of four different Y-shaped configurations of the telescopes feature nine dishes on each arm. By reconfiguring the telescopes, the resolution of the radio images can be adjusted. The array remains in the same configuration for about four months, then the dishes are rearranged into another configuration. In this way, the array cycles through the four configurations in about 16 months. At the maximum resolution, the array acts like a single gigantic telescope 22 miles across with the ability to pick out a golf ball from a distance of 100 miles. The VLA has been used to detect water on Mercury, map the center of the Milky Way Galaxy, and examine black holes.

The Very Large Array in New Mexico.

Visiting Information

At the Visitor Center, start with the theater's nine-minute video, which provides an introduction to radio astronomy, interferometry, and the VLA. Another video shows how the antennas are moved. Other exhibits tell about current research conducted with the VLA. The VLA visitor center is open daily from 8:30 A.M. until sunset. There is no admission charge, and no food service is available although there are soft drink vending machines. At the back door, pick up a brochure that guides you on a walking tour. First stop is the Whisper Gallery, a set of parabolic reflectors placed far apart to demonstrate how the dishes reflect and focus radio waves. One person quietly whispers into one of the dishes and a friend can hear the whisper by listening at the focal point of the other. Now it's on to the base of one of the giant dishes. Climb to the observation deck for a great view of the entire array. When you get back to your car, turn right out of the visitor center parking lot and drive toward the tall building, following the Antenna Assembly Building (AAB) Tour signs. After crossing the railroad tracks, turn left into a parking lot. There you see an antenna undergoing maintenance and one of the transporters used to move the antennas. Quarterly guided tours are offered free of charge to the general public, and during the summer students provide free tours on the weekend.

To get to the VLA from Albuquerque, take I-25 to Socorro. The exit for the VLA is well marked. When you exit the interstate, watch carefully for the sign for U.S. 60 West toward Magdalena. U.S. 60 is Spring Street in Socorro. There is no "to VLA" sign. Continue on U.S. 60 for

> Website: www.aoc.nrao.edu
> Telephone: 505–835–7243

approximately 50 miles. You will see the radio telescopes off in the distance. Just follow the VLA signs.

4
Planetaria—
Theaters of the Sky

Never before was an instrument created which is so instructive as this;
never before one so bewitching; and never before did an instrument speak
so directly to the beholder. The machine itself is precious and aristocratic. . . .
The planetarium is school, theater, and cinema in one classroom under the
eternal dome of the sky.   Elis Stromgren, astronomer

The modern planetarium is the culmination of a 2,000-year-old dream to capture the majesty of the night sky and bring it down to Earth. The first mortal to attempt this feat was the Greek scientist Archimedes who, in 250 B.C., constructed a mechanical device that demonstrated the motions of the Sun, Moon, the Earth, and the planets. During the Middle Ages, the Italian astronomer, astrologer, and mathematician Johannes Campanus (1220–1296), described a mechanical sky model in his book *Theorica Plane-tarum* and was kind enough to include instructions on how to build one. Of course, these devices were not planetaria in the modern sense of the word; they are what we now call orreries, named after the town of Orrery in Ireland where an earl had one built in the 1700s.

The first step in the direction of modern planetaria was the construction of giant hollow spheres in which you could actually stand. The most famous of these primitive planetaria is the Great Gottorp Globe, constructed in the mid-seventeenth century by the German scholar Adam Olearius. The Gottorp Globe, later given to Peter the Great, is now on display at the Museum of Anthropology and Ethnography (the Kunstkammer) in St. Petersburg, Russia. The globe measures about ten feet across, weighs more than three tons, and has a hole through which people could enter and sit on a circular bench inside. A map of the Earth is drawn on the outside surface while the inside surface is painted with a map of the sky. In later designs of this type, small holes were cut in the globe allowing light to shine in to imitate the stars. One of the last of these sky globes was built in 1913 for the Museum of the Chicago Academy of Science by Charles Atwood. The Atwood Sphere has a diameter of 15 feet, shows 692 stars, and features a moveable light bulb that functions as the Sun. A series of openings can be covered or uncovered to represent the planets. The Atwood Sphere is on display at the Adler Planetarium in Chicago where you can even ride a lift to take you up to the inside of the sphere.

The story of the modern projection-type planetarium begins with Max Wolf, a German astronomer who helped establish the Deutches Museum in Munich, a new institution devoted to science and technology. The driving force behind the museum was engineer Oskar von Miller, who had founded the museum in 1903 with the help of several well-known German scientists. In 1913, Wolf approached von Miller with the idea of somehow realistically reproducing the night sky in all its glorious detail. Von Miller liked the idea and discussed it with the highly regarded optical firm of Carl Zeiss located in the German city of Jena. The company agreed to explore the problem. The original idea was to build an improved version of a giant globe like the Gottorp, but in 1919, Walther Bauersfeld, the firm's chief design engineer who later led the company, came up with the idea of optically projecting lights representing stars and other celestial objects onto the ceiling of a dark room. A projection device had several advantages over a globe, including the fact that it was much smaller and more easily controlled.

For five years, Bauersfeld and a large staff of scientists, engineers, and draftsmen struggled with the immensely complicated problem. They had to figure out how to mimic the daily and yearly motions of the stars and planets with optical and mechanical systems and how to interconnect these systems so that the planets would remain in the correct positions with respect to each other.

Finally, in August 1923, a 52-foot-diameter dome was set up on the roof of the factory, and scientists successfully tested the very first Model I Zeiss projector. The projector basically consisted of a hollow sphere with a light inside—a so-called "star ball." Small pinholes in the sphere allowed a nar-

A typical dumbbell-shaped planetarium projector with two "star balls" at either end.

row beam of light to escape—the brighter the star, the bigger the pinhole. When the beam hit the ceiling it made a small point of light that represented a star. The star ball was mounted so that it could rotate to simulate the rotation of the Earth. Later projectors often had two star balls mounted at either end of the projector in a configuration reminiscent of a dumbbell. One ball projected the northern hemisphere stars and the other, the south. The Zeiss projector, known as the "Wonder of Jena," was shipped to the Deutches Museum where the projector had its first public showing. The response was enthusiastic, the word spread, and soon cities all across Europe were ordering projectors.

In 1928, the Chicago philanthropist Max Adler heard of the "Wonder of Jena" and took his wife and an architect all the way to Germany to see it. Adler decided to buy one for his hometown, and in 1930 the Adler Planetarium, the first planetarium in the United States, opened its doors. Samuel Fels of Philadelphia and Charles Hayden of New York quickly followed suit and donated projectors to their cities. The Fels Planetarium in Philadelphia was the second to open in the United States followed by the Griffith Observatory's Planetarium in Los Angeles and New York's Hayden Planetarium in 1935.

Not surprisingly, during World War II, the Zeiss Company produced very few planetaria. After the war, Soviet troops occupied Jena and dismantled the factory. However, before the occupation, the Allies managed to sneak out about a hundred of the company's top personnel. These refugees reorganized, built a new factory in the West German town of Oberkochen, and began producing planetaria in 1956. Meanwhile in Jena, the remaining employees there also reorganized and rebuilt the factory. During the Cold War, two Zeiss companies operated: one in East Germany and the other in West Germany. When Germany reunified in 1989, so did the two Zeiss firms. Today, the Carl Zeiss Company remains the world's leading manufacturer of top-of-the-line planetaria.

In 1936, a Philadelphia newspaperman named Armand Spitz landed a part-time job as a lecturer at the new Fels Planetarium. He quickly realized the planetarium's potential as a teaching tool for schools, colleges, and small museums. The problem was that the Zeiss projectors were much too expensive for schools to buy. So Spitz decided to build a projector that would produce an adequate, but much less expensive, reproduction of the sky. The first Spitz projector, shaped like a dodecahedron (a solid composed of 12 pentagon-shaped faces) rather than a sphere to save on machining costs, was demonstrated to an audience of astronomers in the late 1940s. As a result of his efforts to design an affordable planetarium, Spitz became known as the Henry Ford of the planetarium field. With the advent of *Sputnik* and the concern it brought regarding the quality of science and mathematics education in the United States, hundreds of Spitz planetariums were installed in high schools and colleges across the country from 1964 through the 1980s.

Today, an increasing number of planetariums are turning to digital technology to replace their traditional mechanical projectors. The digital systems have fewer moving parts and do not require the degree of synchronization that traditional systems need. This results in lower maintenance costs and greater reliability. In a fully digital planetarium, an image of the night sky is generated by computer and projected onto the dome using a laser, LCD, or some other type of projector. The projection system can consist of either a single projector in the center of the dome that uses a fish-eye lens to spread the image out over the dome or several projectors located around the rim of the dome each responsible for its own slice of the sky. Modern planetaria can perform amazing tricks like showing the sky as it would look at any latitude on Earth or at any time in the past or present.

They can even dazzle the audience by showing how the positions of the stars would shift while traveling through space.

There are literally hundreds of public planetaria scattered across the United States. Nearly every major city has at least one. Many planetaria have astronomy-related exhibits that supplement the planetarium theater experience, making them more like astronomy museums. A few planetaria have public observatories and telescopes at the ready inviting you to take a peek. A list of the world's planetaria, sorted by country and city, is available on the International Planetarium Society's website at www.ips-planetarium.org. A scroll down the list reveals that the United States has far more planetaria than any other country. In fact, roughly 40 percent of the world's planetaria are located in the United States. Descriptions of the planetaria in the three largest U.S. cities—New York, Los Angeles, and Chicago—are given below. These are arguably the leading planetaria in the country.

Rose Center for Earth and Space, American Museum of Natural History, New York City

The Rose Center for Earth and Space is nothing short of a work of art. The building, featuring an 87-foot-diameter metal sphere enclosed in an enormous glass box 95 feet tall and 124 feet wide, is architecturally magnificent. The grandeur of the building is best appreciated from outside, so be sure to exit the building and take a look. Better yet, come back at night when the building is bathed in light. The stone base of the building's exterior is covered with gray "jet-mist" diorite from Virginia. This stone contains flecks of white feldspar that produce patterns reminiscent of nebula. The enclosed sphere symbolizes stars, planets, and other objects in the universe that assume a spherical shape as a result of the inexorable inward pull of gravity. The upper half of the sphere holds the Hayden Planetarium's Space Theater, while the bottom half cradles the Big Bang Theater. Diffraction gratings take advantage of sunlight shining through the glass to produce rainbows of color that dance on the center's walls and floors. The floors of black composite-stone are embedded with Czechoslovakian glass shards evoking the stars in the blackness of space.

Consider beginning your visit with the "Size Scales of the Universe" exhibit on the top level, take in the short presentation in the "Big Bang Theater," then continue down the spiral "Cosmic Pathway" leading to the "Hall

Scale models of the planets at the Rose Center in New York. The Hayden
Sphere can be seen in the upper left with Jupiter nearby.

of the Universe" on the lower level, and finish with a show in the Space Theater. Oh, and don't forget the gift shop!

The central themes of the Rose Center are space and time, and the two
main exhibits explore these concepts in detail. The best exhibit is the "Size
Scales of the Universe," a square walkway around the center of the sphere
that compares the size of objects in the universe. Whereas most museum displays dealing with the scale of things stick to the relative sizes of the Sun and
the planets, this exhibit does that and much more. In a journey reminiscent
of Charles Eames's classic film *Powers of Ten* the exhibit takes the visitor on
a journey from the very largest scale to the very smallest; the entire universe,
down to a subatomic particle, spans a size range from 10^{24} meters all the way
down to 10^{-18} meters. The display cleverly uses the Hayden sphere as a reference for its comparisons. For example, one panel reads: "If the Hayden
Sphere is the size of the supergiant star Rigel, then this model shows the size
of the Sun." At the other end of the scale another panel reads: "If the Hayden Sphere is the size of a virus, the model is the size of a hydrogen atom
and the models hanging above are molecules of water, ammonia, and
methane." This exhibit is arguably the world's best presentation on the relative sizes of objects.

At the end of the "Size Scales of the Universe" you come to three panels that discuss the Big Bang Theory and the evidence supporting it. At the end of these panels you see the entrance to the Big Bang Theater, housed in the lower half of the sphere. (This is free, and no ticket is required.) Go inside the theater and watch the short four-minute presentation narrated by Maya Angelou. The Big Bang marked the beginning of time, and the Big Bang Theater is the starting point for your walk through time. Upon exiting the theater, you begin your descent down the 360-foot downward sloping "Cosmic Pathway," that spirals around the Hayden Sphere, taking you on a journey through 13 billion years of cosmic history. Each foot along the pathway moves you forward in time by 45 million years, and each meter is equivalent to 147 million years. A scale on the pathway lets you measure how many years each of your steps takes you through. The path is marked by panels that inform you of significant cosmic events that take place as you move forward through time.

After your journey through space and time, you are now ready to explore the Hall of the Universe. Here, you find informative panels on various astronomical topics. The floor is littered with digital scales that reveal your weight on Mars, the Sun, a red giant star, and other astronomical bodies. Be sure to check your weight on a neutron star! The Astro-Event Wall is a giant video screen where you can watch eight- to ten-minute astronomy news segments.

It's hard to miss the Willamette meteorite, found near what is now West Linn, Oregon, in 1902. This iron meteorite, weighing in at 15.5 tons, is the largest meteorite ever found in the United States and is one of the largest in the world. Notice the large, irregular cavities in the top side of the meteorite. These holes were formed by rainwater combining with iron sulfide deposits in the meteorite to form sulfuric acid which, over thousands of years, gradually ate away at the metal. To see more meteorites, visit the Museum's Author Ross Hall of Meteorites.

Artistically inclined scientific travelers should be sure to visit the Weston Pavilion where an 18-foot, 3,500-pound sculpture of an armillary sphere hangs from the ceiling. At the center of the aluminum and stainless steel sphere is an S-shaped piece of metal representing the spiral arms of our home galaxy, the Milky Way. The sphere is oriented so that it indicates the exact location of New York City on January 1, 2000.

The Rose Center also includes the Hall of Planet Earth, which holds an array of geology exhibits concerning plate tectonics, earthquakes, volcanoes, and the formation of planet Earth. A working seismograph is on display along with simulated ice core samples.

Of course, no visit to the Rose Center would be complete without taking in a show at the 429-seat Space Theater. This theater is advertised as the world's largest, most powerful, and highest-resolution virtual-reality facility. At the center of the dome is the four-ton Zeiss Mark IX Hayden Edition star projector, a one-of-a-kind instrument that creates an image of the night sky with more than 9,000 stars on the theater dome. The projector can even make the stars twinkle.

Visiting Information

The Rose Center for Earth and Space is part of the American Museum of Natural History located at Seventy-ninth Street and Central Park West in New York City. The main entrance to the museum is on Central Park West, but there is an entrance on the north side of the museum on West Eighty-first Street that leads directly into the Rose Center. The easiest way to get to the museum is to take the B or C subway line to Eighty-first Street. There is an entrance to the museum at the subway stop, and the ticket line there is usually shorter than the lines at the main entrance. The museum is open every day from 10:00 A.M. to 5:45 P.M., except Thanksgiving and Christmas. On the first Friday of every month, the museum stays open until 8:45 P.M., and you can listen to live jazz in the evening. General admission to the museum is $14 for adults, $8 for children ages 2 to 12, and $10.50 for senior citizens and students with identification. If you want to see a show in the Space Theater, buy a combination ticket for $22 for adults, $13 for children, and $16.50 for seniors and students. You can avoid ticket lines altogether by purchasing your tickets online. Space Shows often sell out on busy weekends and during the summer so it is advisable to get your tickets in advance.

A 75-minute audio tour of the Rose Center, available in English or Spanish, is free with admission. The audio tour guides you along a 34-stop route through the Rose Center and is narrated by author and astronomer Neil deGrasse Tyson, director of the Hayden Planetarium. The format of the audio tour allows you the freedom to tailor the tour to your particular interest. The audio tour devices can be picked up at the desk on the lower level.

Website: www.amnh.org/rose/index.php
Telephone: 212–769–5100

Dining options include the Museum Food Court, which is directly accessible from the lower level of the Rose Center, the Café Pho, which serves Vietnamese cuisine in the Seventy-seventh Street lobby, and the Café on 4, serving light fare while providing a view of the museum grounds.

Griffith Observatory, Los Angeles, California

First, a word of explanation: The Griffith "Observatory" is a public observatory that includes a planetarium and astronomy exhibits along with telescopes for public viewing; it is not a working scientific observatory where professional astronomers come to make observations.

The Griffith Observatory sits on the south-facing slope of Mount Hollywood, just a few miles from downtown Los Angeles. On a clear day, the observatory offers a spectacular view of the city and many people come here just to soak up the sun and take in the scenery. The observatory has served as the setting for scenes in dozens of movies including the original *Terminator*, Disney's *Rocketeer*, and, most famously, as the backdrop for the knife fight scene in *Rebel Without a Cause*. A bust of that movie's star, James Dean, sits just west of the observatory. Observatory director Dr. E. C. Krupp jokes, "We've been in so many movies we ought to have a star on Hollywood Boulevard."

The observatory is named after its benefactor, Griffith J. Griffith, a tycoon who made his millions in Mexican silver mines and California real estate. After touring abroad and admiring the great public parks he encountered in the cities of Europe, he decided that his own home town deserved a public park. So in 1896, he gave some of his land to the city of Los Angles for that purpose. Spread out over 4,210 acres, Griffith Park is one of the largest public parks in the United States. Later in his life, Griffith visited the Mount Wilson Observatory, looked through the 60-inch telescope, and was awestruck. Griffith became convinced that looking at distant heavenly objects would give people a new, more enlightened perspective. According to Griffith, "Man's sense of values ought to be revised. If all mankind could look through that telescope, it would change the world!" Griffith decided to give the city $100,000 to build an observatory on the top of Mount Hollywood where the public could come to gaze at the stars—not the kind in the movies, but in the sky. Griffith, who died in 1919, did not live to see his vision realized; however, the Griffith Trust continued his work and, in 1930, enlisted the help of some leading astronomers of the day to help design the observatory. None other than George Ellery Hale steered the overall design, and Russell Porter, the "patron saint" of the amateur telescope-making movement, provided valuable advice. Because it was built during the Great

Peter Weber/Shutterstock

The newly renovated Griffith Observatory in Los Angeles.

Depression, the builders took advantage of low prices and used the best materials available; the result is a building both durable and beautiful. Six sculptors from the depression era Work Projects Administration created the Astronomers Monument located on the lawn. The building opened with great fanfare in 1935, and the city of Los Angeles acquired ownership.

The Griffith Observatory has recently been given a $93-million facelift that includes a modernization of the planetarium, 60 new astronomy exhibits, and 40,000 square feet of added public space. The brass, concrete, and copper exterior has been restored to Russell Porter's original design. The renovation and expansion has returned the Griffith Observatory to the ranks of a world-class astronomy education facility.

The heart of the observatory is the 300-seat Samuel Oschin Planetarium, newly equipped with a seamless dome, a laser digital projection system, a theatrical sound and lighting system, and a Zeiss Universarium Mark IX Star Projector, the most advanced instrument of its kind in the world. In contrast to nearly all other major planetaria, this planetarium continues its tradition of having a live narrator present the planetarium program. In addition to the planetarium, the observatory boasts a new 200-seat multimedia theater dubbed the Leonard Nimoy Event Horizon in honor of the actor who made a large financial contribution to the building of the theater. The theater, used for lectures, demonstrations, films, serves as a major venue for its school and educator programs.

The Griffith Observatory has several high-quality telescopes that stand ready to give the public a peek at the cosmos. The observatory's main telescope located in the east dome is a 12-inch Zeiss refractor, a fixture there since the observatory opened. More than seven million people have put their eye to this telescope; in fact, more people have looked through this telescope than any other telescope in the world. The main telescope tube is 16 feet long and carries a nine-inch telescope on its back. The total weight of the telescope is a hefty 9,000 pounds. To prevent the telescope from bending under its own weight, a system of counterweighted levers is built into the tube and mounting. If you visit during the day, be sure to climb up on the roof and take a look at the telescope. On clear nights, you can actually look through it. The observatory staff also sets up telescopes on the lawn or the East Observatory Terrace for additional public viewing. Every month, the Los Angeles Astronomical Society and Sidewalk Astronomers host a star party; visitors bring their telescopes, and the public is invited to take a look.

As you enter the building's main entrance, linger a while at the Central Rotunda and watch one of the world's largest Foucault Pendulums swing back and forth. As the Earth rotates beneath it, the pendulum knocks over pegs that stand in the pendulum pit. Be sure to admire the beautiful murals painted by Hugo Ballin. The ceiling murals represent celestial mythology while the rectangular wall murals celebrate the advancement of science in eight areas: astronomy, aeronautics, navigation, civil engineering, electricity and metallurgy, time, geology and biology, and mathematics and physics.

The Griffith Observatory has four main exhibit areas: The Hall of the Eye, The Hall of the Sky, The Edge of Space, and The Gunther Depths of Space. The Hall of the Eye traces the progress that humans have made in observing the sky and describes instruments like the telescope that improve upon naked eye observations. On your left as you enter the hall, take a look at the Camera Obscura, an instrument that uses mirrors and lens to focus light onto a flat surface. This one uses a periscopelike tube mounted on the roof to produce a 360° view of the Los Angeles basin on a large circular disc. Through dioramas the "Using the Sky" exhibit illustrates how ancient peoples used the sky to gain some degree of control over their lives. The center of the hall features an optical workbench where you can play around with mirrors and lenses to understand the operation of a telescope. On one side of the central exhibit is a copy of Edwin Hubble's photographic plate in which he discovered a variable star in the Andromeda Nebula. In his excitement, Hubble writes "VAR!" on the plate. The "Observing in California" exhibit uses photographs, models, and artifacts to describe the leading role California observatories played in late nineteenth- and twentieth-century astronomy. Here, you can see the mirror plug from the Palomar 200-inch telescope. For the "Beyond the Visible" exhibit a large glass wall display shows how the various parts of the electromagnetic spectrum open new windows to the universe. Finally, do not miss the Tesla Coil, on display here since 1937. This device, an invention of the electrical wizard Nicola Tesla, can throw long electric sparks through the air.

The Hall of the Sky spotlights the two most familiar objects in the sky: the Sun and the Moon. Here, you find the solar telescopes, models, and animation illustrating the properties of the Sun and kinetic displays on how the motions of the Sun, Moon, and the Earth produce night and day, the seasons, the phases of the Moon, tides, and eclipses. A giant periodic table of the chemical elements holds samples of most elements and relays one of the most amazing facts of science: all of these elements, except for hydrogen, are made by stars. Three solar telescopes, located in the Hall of the Sky, give visitors a chance to view the Sun. A triple-mirrored tracking device, called a coelostat, perched high above in the western dome, directs beams of sunlight into the three telescopes. One telescope shows an image of the Sun in ordinary white light, and another provides a view through a hydrogen-alpha filter, while the third reveals the solar spectrum.

The corridor connecting the original building to the underground exhibits, playfully called the "Cosmic Connection," features a 150-foot timeline of the universe with significant cosmic events highlighted with images and decorated with thousands of pieces of jewelry. Notice that all human history occupies only a fraction of an inch.

The corridor leads you down to the "Edge of Space" exhibits where you can examine meteorites, including lunar and Martian meteorites, fragments of the meteorite that created the meteor crater in Arizona, and a meteorite from the asteroid Vesta. The neighboring asteroid impact simulator is fun to play with. At the center of the mezzanine, a cloud chamber allows you to actually see the path of cosmic rays coming in from outer space. The spark chamber discharges every time a cosmic ray hits its electrically charged surface. Nearby a large globe of the moon accurately portrays its mountains, craters, and maria. An Apollo Moon rock accompanies the globe.

After catching a film in the Nimoy Event Horizon Theater, follow the ramp down into the new Gunther Depths of Space exhibit beneath the original building. There, the observatory's most visually stunning display, appropriately dubbed "The Big Picture," greets you. At 152 feet long and 20 feet high, this is the largest single astronomically accurate image ever made. Astronomers pieced it together from observations made with the 48-inch telescope at the Palomar Observatory as part of the Palomar-Quest astronomical sky survey. More than a million individual stars, galaxies, and other celestial objects are visible in the image. As you move toward the Big Picture, you may notice that your perception of the objects changes. What you thought was a star from far away might, as you move closer to the image, turn out to be a giant spiral galaxy. Ironically, The Big Picture only shows a tiny slice of the sky in the constellation of Virgo—a slice about the size of what your index finger would cover if you held it a foot in front of your eye. If you look at this small parcel of sky with your naked eye, you see only a few lonely stars. Across from The Big Picture a bronze statue of Einstein, sitting on a bench with his index finger extended, illustrates the point.

Moving into the exhibit, you find scale models of the planets mounted on tall pylons surrounded by informative panels. On this scale, the Nimoy Theater represents the Sun. Bathroom scales rest in the floor so you can get a feel for each planet's gravity. If you look up and to the left of the planet models, you see a projected model of the solar system that gives an accurate representation of the relative motions of the planets. At the "Other Worlds, Other Stars" station, you can learn all about the discovery of planets outside our own solar system, including what these new systems might look like. The Milky Way Galaxy display uses an eight-foot-diameter glass model of our galaxy to illustrate its three-dimensional shape. Recent images of distant celestial objects are shown on a nearby giant screen that also serves to bring visitors up-to-date on the most recent astronomical news. Tucked away into a corner is the previous Zeiss Planetarium Projector that was used here from 1964 through 2002.

Visiting Information

Griffith Observatory is located in Griffith Park near Hollywood in Los Angeles. The observatory is open Tuesday through Friday from noon until 10 P.M. and on Saturday and Sunday from 10 A.M. until 10 P.M. The observatory is closed on Mondays, Thanksgiving Day, and Christmas Day. A night visit enables you to look through the telescope and gaze at the twinkling lights of Los Angeles. Admission to the observatory and exhibits is free; however, there is a charge for shows in the Oschin Planetarium. Ticket prices are $7 for adults and children 13 and older, $5 for seniors, and $3 for children ages 5 to 12. Children 4 and younger may only attend the first planetarium show each day; they receive free admission, but they must sit on the lap of a parent or guardian. The Stellar Emporium Gift

> Website: www.griffithobs.org
> Telephone: 213–473–0800

Shop has all sorts of astronomy-related items to spend your hard-earned money on, and if you get hungry, The Café at the End of the Universe, operated by Wolfgang Puck, can satisfy your appetite.

Adler Planetarium and Astronomy Museum, Chicago, Illinois

The Adler Planetarium sits majestically at the end of a half-mile-long peninsula that juts out into Lake Michigan and offers a spectacular view of Chicago's skyline. The building itself is a juxtaposition of old and new: a modern semicircle of glass surrounds the original 12-sided art deco building clad in rose, gray, and purple rainbow granite. The "Sky Pavilion," added in 1999, greatly expanded the facility. Several noteworthy artistic works adorn the grounds. Directly in front of the building is a statue of Nicolaus Copernicus, the Polish astronomer who first proposed that the Sun lay at the center of the solar system. Erected in 1973 to celebrate the 500th anniversary of Copernicus's birth, this is a copy of a statue that sits outside the Polish Academy of Sciences in Warsaw. Notice the armillary sphere in Copernicus's left hand with, presumably, the Sun at the center. On your left as you face the building is a 13-foot sundial designed by British sculptor Henry Moore. The thin rod held by the larger vertical crescent casts a shadow on the time scale on the horizontal crescent; this particular form of this sundial is considered the most accurate shape for this instrument. To your right is a shiny stainless steel sculpture called "Spiral Galaxy." Unlike some modern sculptures, this one actually looks like what it's supposed to represent. Behind the building and to the right is a sculpture called "America's Courtyard." This installation has multiple levels of symbolism: the concentric circles evoke the pattern at Stonehenge and other archeological sites; the increasing height of the stones as you go out from the center suggest an amphitheater, and the multiple colors and kinds of stone represent the multicultural melting pot that is America. Pathways through the stones

The Adler Planetarium in Chicago.

mark the points on the horizon where the rises and sets on the summer and winter solstices.

Directly behind the main building is the Doane Observatory, home to a 20-inch reflecting telescope. On "Far Out Fridays," held on the first Friday of every month, the observatory is open to the public for astronomical viewing. The planetarium also offers a lecture series and a science fiction film series.

Upon entering the building, you will find yourself in the "Rainbow Lobby," so named because the afternoon sunlight shining through the beveled glass in the entrance doors produces dozens of little rainbows on the walls. Across from the entrance are bronze emblems for each of the eight planets known at the time the plaques were installed. Pluto was discovered only five weeks before the Adler opened in 1930, too late for a plaque. With the demotion of Pluto to the status of a dwarf planet, it turns out that eight plaques are enough! The museum part of the Adler Planetarium is divided into galleries on the upper and lower levels. The main galleries on the upper level focus on the solar system and the Milky Way galaxy. In the Solar System gallery, you can see a model of the Mars Rover and then try your hand at maneuvering the remote-controlled robotic "Adler Rover." A large plastic globe filled with a blue liquid is used to represent the outer gas giant planets. Spinning the globe illustrates how the banded atmospheric structure of these planets is formed. Another fun exhibit is the "Make an Impact" demonstration where a ball is shot into sand to simulate the formation of craters. The gallery also features the obligatory scale models of the planets. Look up to see the giant Jupiter, Saturn, Uranus, and Neptune models suspended from the ceiling. A giant scale model Sun separates the Solar System gallery from the Milky Way gallery. Here, you find exhibits explaining the life cycle of stars and describing the internal structure of the Sun. At the black hole exhibit, you can release balls that spiral down and disappear into the "gravity well." Did you ever wonder what would happen if you jumped feet-first into a black hole? The tremendous gravitational tidal forces would stretch your body out into human spaghetti, a process astronomers have appropriately dubbed "spagettification." Check yourself out in the curved carnival-like mirrors that painlessly reproduce this effect on your body. In a small mini-theater at the gallery entrance, glasses allow you to view images of Mars and the Milky Way galaxy in three dimensions.

In the lower galleries, we come to what sets the Adler Planetarium apart from other planetaria. When he founded the planetarium in 1930, Max Adler bought a superb collection of about 500 astronomical, mathematical, and navigational instruments from a collector in Europe. Today, the collection has grown to nearly 2,000 instruments, including astrolabes, armillary spheres, celestial globes, nocturnals, orreries, sundials, and telescopes dating from the twelfth through the twentieth centuries. This instrument collection is the largest of its kind in the western hemisphere and one of the largest in the world. The Adler rare book collection houses more than 3,000 volumes. A small fraction of the collection is on permanent display in the lower galleries, especially in "The Universe in Your Hands" gallery. To allow the public to see a bit more of the collection, the Adler offers a temporary exhibit series called "Special Topics in the History of Astronomy." This exhibit, which changes approximately every three months, is organized around various themes, such as Islamic astronomy, Galileo's telescopes, and women in the history of astronomy.

Also in the lower gallery, you find the "From the Night Sky to the Big Bang" gallery, which traces the history of mankind's conceptions of the cosmos from Ptolemy through Hubble. The gallery features the 20-foot-long Dearborn telescope. Built in 1864, it was the largest telescope in the world at the time. At the end of your walk through this gallery, don't miss Alan

Guth's scientific notebook turned to the equation-filled page where he came up with the idea of an inflationary universe.

From Meso-America to Egypt and the Orient, the "Bringing the Heavens to Earth" gallery explores the significance of astronomy in cultures around the world. Here, you discover how the Polynesians used the stars to navigate and how the fate of the Pharaohs was determined by the stars. The many interesting cultural artifacts on display here include a replica of what is perhaps the oldest known (dating from about 30,000 b.c.) astronomical artifact: the bone of an eagle etched with a lunar calendar that was probably used to determine the best nights for hunting and other activities.

Finally, get in line for the lift that will take you up into the belly of the "Atwood Sphere" that was mentioned in the introduction to this chapter. In 1913 Charles Atwood, secretary of the Chicago Academy of Sciences, built the sphere to educate the public about the stars; it is widely considered the predecessor of the modern planetarium. This simple device consists of a fifteen-foot diameter hollow 1/64-inch galvanized iron sphere with 692 small holes drilled in its surface allowing outside light to enter and form the patterns of the constellations. A motor rotates the sphere around the occupants seated on an interior platform in a simulation of the daily rotation of the Earth on its axis.

The newest exhibit at the Adler is called "Shoot for the Moon." The exhibit tells the story of astronaut Jim Lovell—his childhood days spent reading books on rocketry and science fiction, his failure to be accepted into the Mercury program owing to a medical technicality, and his four NASA space missions: *Gemini VII* and *XII* and *Apollo 8* and *13*. Of course, Lovell is most famous as the commander of *Apollo 13* during which an oxygen tank exploded and ripped a hole in the ship. Due to the creative problem-solving ability of the ground crew and Lovell's calm leadership, the ship was returned safely to the Earth. The flight of *Apollo 13* was dramatized in the Academy Award-winning movie *Apollo 13* starring Tom Hanks as Lovell. The exhibit contains nearly thirty artifacts from Lovell's personal collection, including his letter of rejection from the Mercury Program, his handwritten Gemini training notes, and the manuals and flight plans for the *Gemini XII* mission. The centerpiece of the exhibit is the fully restored *Gemini 12* spacecraft, the last mission of the Gemini program flown by Lovell and Buzz Aldrin in 1966.

Of course, no visit to the Adler would be complete without seeing the traditional planetarium show. The Adler boasts that it is the only museum in the world with two full-size planetarium theaters: the original Sky Theater and the recently added Star Ride Theater. Several different shows are available throughout the day, and, depending on the option you choose, your admission can include one or two of the shows.

Visiting Information

The Adler Planetarium is located in downtown Chicago at 1300 South Lake Shore Drive. Parking is available on the museum campus for $15 per day. If you're staying at a downtown hotel, you may want to consider taking a taxi or mass transit. The museum is open daily from 9:30 A.M. to 4:30 P.M. with extended hours in the summer. Admission to the museum only is $7 for adults, but there are a couple of admission packages to choose from that include shows and an audio tour device. The packages start at $16 for adults. If you are visiting other Chicago attractions, you may want to consider buying a Museum Campus Pass, a CityPass, or a Go Chicago card. These options can save you considerable money on admissions. Food is available in Galileo's Café with large glass win-

Website: ww.adlerplanetarium.org/
Telephone: 312–922–782

dows looking out over downtown Chicago and Lake Michigan. The Infinity Shop has a wide variety of astronomy-related items.

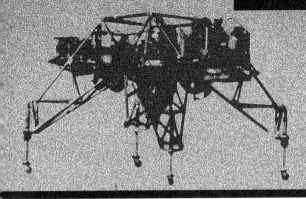

5
NASA and Space Exploration

Space, the final frontier. . . .

Captain James T. Kirk
of the Starship *Enterprise*

When future historians look back on our present age, they will remember it primarily as the time that our species tentatively dipped its toes into the great cosmic ocean and began the exploration of space. The final frontier was opened by the advent of rockets. The basic scientific principle behind the operation of a rocket is Newton's Third Law of Motion: For every action, there is an opposite and equal reaction. In other words, if Object A exerts a force on Object B, then Object B exerts an equal force in the opposite direction on Object A. To illustrate this principle, consider blowing up a balloon with air and releasing it. The balloon whizzes wildly around the room. Let Object A be the balloon and Object B be the air inside the balloon. When the balloon is released, the stretched-out rubber of the balloon (Object A) exerts a force on the air inside (Object B), pushing the air out. The reaction is that the escaping air (Object B) exerts an equal force in the opposite direction on the balloon (Object A) propelling it forward. Action: Balloon exerts a force on the air. Reaction: Air exerts an equal and opposite force on the balloon. A rocket works basically the same way, except that the air is replaced by combustible fuel and the rubber balloon is replaced by the metal body of the rocket. The force the combustible fuel exerts on the rockets is called the thrust.

NASA

Dr. Robert Goddard uses a Ford Model-A truck to tow a rocket to the launching tower located outside of Roswell, New Mexico, in the early 1930s.

A common misconception is that the fiery gases escaping the nozzle of the rocket engine must have some material external to the rocket itself, the concrete launching pad say, or the surrounding air, to push against. In fact, when Robert Goddard first proposed that rockets could be used for traveling in outer space, he was ridiculed by some journalists who argued that rockets could not possibly work in the vacuum of space because there was nothing out there to push against. Goddard proved his critics wrong by showing them a demonstration in which a rifle was suspended by wires in a glass enclosed vacuum chamber. When the rifle was fired, the bullet went one way, and the rifle kicked back in the opposite direction; thus, he proved that the bullet didn't need air or anything else to push against to propel the rifle back in the opposite direction.

While Newton's Third Law makes a rocket work, his First Law makes a rocket practical. Newton's First Law states that if no unbalanced force acts on an object, then an object at rest will stay at rest and an object moving at a constant velocity (that is, in a straight line at a constant speed), will continue to move at that constant velocity. Most people have no trouble understanding and accepting the fact that an object at rest will stay at rest. This agrees with our everyday experience: Objects at rest don't just start moving on their own; a push or a pull is required to move them. The part of the law that is much more difficult for people to accept is the statement that an object moving at a constant velocity will continue this motion. The problem is that this doesn't agree with our everyday experience. To keep an object moving, we have to keep pushing. But this is true because here on the surface of the Earth, frictional forces, like air resistance, are always present. If there were absolutely no frictional or other unbalanced forces acting on an object, then once the object is set into motion, it will continue to move at that exact speed, in that exact direction, forever. This is what happens in space. There's no air in space, so there's no air resistance. To get a rocket from point A to point B in space, the rocket engines are fired to attain the desired speed and point the rocket in the desired direction. Then, the rockets are turned off, and the rocket continues to move at that speed in that direction. In space, you don't have to keep pushing by continually firing the rockets. If you did, the rocket would have to carry a prodigious amount of fuel—enough to make a space trip of any significant distance utterly impractical.

The Chinese invented rockets fueled by gunpowder in the 1300s and used them for ceremonies and celebrations. In the early 1900s, the American physicist and rocket pioneer Robert Goddard launched the first liquid propellant rocket for high-altitude flight and dreamed of one day sending rockets into outer space. During World War II, German rocket scientists, building on the work of Goddard, developed the V-2 rockets that rained down on England. After the war, the leading German rocket scientist, Wernher von Braun, emigrated to the United States and developed rockets

Dr. Werner von Braun in his office at the Marshall Space Flight Center in 1960.

for military applications and eventually, for space exploration. Meanwhile, in the Soviet Union, the rocket development program, under the leadership of Sergei Korolov, kept pace with and eventually surpassed the U.S. efforts.

What became known as the space race began with the launching of the Russian satellite *Sputnik* in October 1957. The shocked United States quickly answered in early 1958 with *Explorer I*, which made the first scientific discovery in space, the Van Allen radiation belt. To direct U.S. efforts in space, a new government agency called the National Aeronautics and Space Administration (NASA) was established. NASA took over four laboratories that had been used by the National Advisory Committee for Aeronautics (NACA) and began operations in October 1958.

In May 1961, President John F. Kennedy boldly defined the direction of the U.S. space program by issuing the following challenge: "I believe that this nation should commit itself to achieving the goal, before this decade is out, of landing a man on the Moon and returning him safely to the Earth." Kennedy's goal of putting a man on the Moon was more about Cold War politics and prestige than about pure science. The United States had been embarrassed by a series of Soviet "firsts" in space including the first person in space, Yuri Gagarin, who orbited the Earth in 1961. Moreover, the Kennedy administration had been humiliated by the Bay of Pigs fiasco. Kennedy had been advised that going to Moon was one aspect of space exploration where the United States actually had a lead over the Soviets and that we could probably beat them to it.

The giant leap to the Moon was broken up into three smaller steps: Mercury, Gemini, and Apollo. The Mercury program was a series of unmanned and six one-man missions with three main objectives: (1) to test a human's ability to function in space; (2) to successfully place a manned spacecraft in orbit around the Earth; and (3) to safely recover the astronaut and spacecraft after the mission. The original *Mercury 7* astronauts—Scott Carpenter, Gordon Cooper, John Glenn, Gus Grissom, Walter Schirra, Alan Shepard, and Deke Slayton—were carefully chosen military pilots. In 1961 Alan Shepard became the first American in space, and in 1962 John Glenn became the first American to orbit the Earth.

The second U.S. manned spaceflight program involved a two-man crew; hence, the name Gemini from the constellation Gemini, the Twins, was adopted. The Gemini Program consisted of ten manned missions from 1965 through 1966. The Gemini Program had three goals: (1) to subject men and equipment to space flights of up to two weeks in duration; (2) to rendezvous

NASA

This is the boot print of Buzz Aldrin made on the surface of the moon on July 20, 1969.

and dock with orbiting vehicles and to maneuver the combined craft; and (3) to perfect methods of reentering the Earth's atmosphere.

The Apollo program, taking the final steps to the Moon, culminated in astronaut Neil Armstrong's first step on the Moon on July 20, 1969. The Apollo missions successfully landed a dozen men on the lunar surface. The last Apollo mission, *Apollo 17*, carried the first scientist, geologist Harrison Schmitt, to the Moon in late 1972. We haven't been back since then. It is difficult to overstate the significance of the Moon landing. It is undeniably the single most important event in the twentieth century and certainly one of the most important benchmarks in all of human history. Think of it: for the first time the human species ventured forth from our biological cradle and set foot on another world.

In the 1970s, both the United States and the Soviet Union established space stations in Earth orbit. The Soviet *Salyut* (Russian for "salute" or "fireworks") space stations were a series of temporary stations first launched in 1971. The *Salyut* laid the groundwork for the highly successful *Mir* (Russian for "peace" or "world") station, the world's first permanent research station in space. First launched in 1986, *Mir* was assembled by connecting modular units together in space. It orbited the Earth for fifteen years during which time cosmonauts consistently broke records for duration in space; some stayed on board for more than four hundred straight days. The first American space station, *Skylab,* was launched in 1973. *Skylab* welcomed three crews of three astronauts in 1973 and 1974. In 1979, owing to technical problems and a lack of funding, *Skylab*'s orbit was allowed to decay and the ship broke apart as it reentered the Earth's atmosphere. Most of *Skylab*'s surviving parts splashed harmlessly into the Indian Ocean, but a few bits and pieces landed in Western Australia, which prompted local Australian authorities to fine the U.S. State Department $400 for littering. The fine went unpaid.

In the post-Apollo era, NASA sought a way to make space travel more frequent and less expensive. They came up with the idea of a reusable Space Shuttle consisting of a reusable orbiter, two reusable solid fuel rockets, and an expendable fuel tank. The orbiter would be attached to one side of the fuel tank with the solid rockets attached to opposite sides of the tank. The Shuttle would be launched into space like a rocket, float back through the atmosphere like a glider, and land on a runway like an airplane. The Shuttle was designed for various purposes, including space-based scientific research and the delivery, servicing, and retrieval of satellites. The military and commercial communities could also use the Shuttle for their space projects.

NASA launched the first operable Space Shuttle, *Columbia,* on April 12, 1981. The Space Shuttle, an undeniably spectacular piece of technology, is beautiful to watch. The Shuttle program has unfortunately been plagued by safety issues. After two dozen safe and successful missions, the twenty-fifth Shuttle mission ended in tragedy when the Space Shuttle *Challenger*'s fuel tank exploded shortly after launch, killing all seven crew members and destroying the orbiter. An investigation revealed that the accident had been caused by the failure of a circular band of rubber called an O-ring that was used to seal the joint between two segments of one solid rocket booster. This failure allowed a flame to escape and ignite the large external fuel tank. NASA investigators determined that the hardening of the rubber in the freezing temperatures the night before the launch probably caused the O-ring failure. This finding raised questions about the decision-making process surrounding the launch.

Therefore, NASA put the Shuttle program on hold for three years while the agency instituted improvements in safety and management. Finally, the

Shuttle blasted off again in 1988. Tragedy struck once more in 2003 when the Shuttle *Columbia* broke apart upon reentering the Earth's atmosphere. This time, the culprit was a chunk of foam that broke off from the external fuel tank during launch and struck the leading edge of the left wing. The impact broke a seal and opened a gap that allowed enough hot gases into the wing during reentry to rip the Shuttle apart.

The Shuttle program took off again in 2005. Currently, three Shuttle orbiters are in service—*Discovery, Atlantis,* and *Endeavour.* They are the only U.S. vehicles available for manned space flight. The Space Shuttle fleet has flown between two and nine missions per year, except after the two accidents. Despite its reuseable design, the Shuttle program has not resulted in more economical space flight. NASA plans to retire the Shuttle fleet by 2010.

So what's the next step in the manned exploration of space? Fusing the best of the Apollo and Shuttle technologies, NASA is currently developing the spaceships that will take us back to the Moon within the next two decades. The Ares I and V rockets will provide the thrust and the Orion Capsule will provide a cradle for the crew. A permanent base on the Moon is scheduled for 2024. This lunar outpost will serve as the steppingstone for an eventual manned mission to Mars.

If measured by the richness of new scientific discoveries, the unmanned space exploration program has far outpaced its manned counterpart. NASA's robotic explorers—with names like *Mariner, Viking, Voyager, Magellan,* and *Galileo*—have ventured to every planet in the solar system and most major moons. The age of planetary exploration began in 1962 when *Mariner 2* (the launch of *Mariner 1* failed) paid a visit to Venus. On its way to Venus, *Mariner 2* made the first scientific discovery in interplanetary space by detecting and measuring the solar wind—a stream of charged particles, mostly protons and electrons, flowing outward from the sun. Arriving at Venus, the *Mariner* spacecraft provided data on the Venusian climate and clouds.

The first successful landing on Mars took place in 1976 when the *Viking 1* and *2* landers touched down on the orange sand. The experiments performed by the *Viking* robots looked for signs of life, but they found none. Recent evidence strongly supporting the idea that Mars was once a warmer and wetter world along with the controversial claim by NASA scientists that a Martian meteorite contains microbial fossils has reignited interest in Mars and made the red planet the primary target for future exploration.

The most ambitious of NASA's robotic explorers were the twin nuclear-powered *Voyager 1* and *2* spacecraft launched in 1977. Both *Voyagers* flew by

Jupiter and Saturn and their moons, and *Voyager 2* added Uranus and Neptune to its planetary itinerary. These ships radioed back to the Earth a mother lode of information about the outer solar system, and their myriad discoveries required the rewriting of astronomy textbooks. The *Voyager* spacecraft are now approaching the boundary of our solar system as they hurtle through the unimaginable vastness of space at a speed of a million miles per day while still in communication with the Earth. Just in case the *Voyagers* have a close encounter with an alien spacecraft, there is a gold-plated record with greetings from Earth in 55 different languages along with some of Earth's greatest musical hits and a variety of earthly sounds and images.

Over the next few years, the Phoenix Mars Scout will land on the polar ice caps and use a robotic arm to dig into the water ice. The Mars Science Lab will roam long distances across Mars collecting and analyzing soil and rock samples and paving the way for a proposed mission to return samples of Martian soil and rock back to the Earth. But Mars is only one of 173 worlds (8 planets and 165 moons) to explore in our solar system so we've only just begun. Plenty of discoveries await, and it is possible that on at least one world, simple life forms will be found.

NASA has a budget of nearly $17 billion (less than one percent of the national budget) and employs more than 18,000 people at its 11 major facilities. All of NASA's facilities have, at the very least, a Visitor Center, although several facilities do not offer public tours. Descriptions of two facilities—the Johnson Space Center in Houston, Texas, and the Marshall Space Flight Center in Huntsville, Alabama—have been deferred until the following chapter because their main attraction is a space museum. The NASA facilities I describe are listed in alphabetical order.

Ames Research Center, Moffett Field, California

NASA's Ames Research Center was founded in 1939 as the NACA's second aircraft research laboratory. The facility, home to 2,300 research personnel, is named in honor of Joseph Ames, who served as the chairman of the NACA for twenty years. Nestled in the heart of Silicon Valley, one natural area of emphasis here is information technology, including networking, artificial intelligence, and advanced supercomputing. Ames is a leader in the field of astrobiology, with a focus on gravity's effects on living organisms and the

An aerial view of wind tunnels at the Ames Research Center. At the lower left is the 12-foot Pressure Wind Tunnel. The giant complex in the center of the photo is the 40- x 80-foot Wind Tunnel, which has been designated as a National Historic Landmark. Just below that is the 14-foot Transonic Wind Tunnel. At the lower right are the twin 7- x 10-foot Wind Tunnels.

nature and distribution of stars, planets, and, possibly, life in the universe. Currently, Ames is heavily involved in NASA's Kepler mission, an orbiting observatory that will search for Earth-sized planets around other worlds. The center also performs research in the areas of nanotechnology, thermal protection systems, and human factors studies. Ames has an impressive array of wind tunnels including the world's largest tunnel.

There are no public tours of the Ames facilities, but visitors are welcome at the "Exploration Center." Exhibits here include the Edgarville Airport, a three-dimensional, interactive display that allows you to manage air traffic at a virtual airport. By inspecting a large topographical globe of the Martian landscape, you can get a feel for the scale of some of Mars's geological features including Olympus Mons, the solar system's largest volcano, and the Valles Marineris, a canyon that dwarfs the Earth's Grand Canyon. Meteorites and a Moon rock are on display; the Moon rock was collected by the *Apollo 15* crew when they landed in the Hadley-Apennine region in 1971. You can also take a look at the Mercury Redstone 1A capsule that was launched in December 1960. This was the last unmanned Mercury test flight before Alan Shepard's first space flight. Notice the porthole that was cracked when the capsule splashed down in the Atlantic Ocean.

The Exploration Center is home to the largest Immersive Theater on the west coast. On the 14-foot-tall, 36-foot-wide curved screen, you can enjoy panoramic views of Saturn and her rings as well as true color images of the Martian landscape. Some images have a resolution twice that of high-definition television.

Visiting Information

The Ames Exploration Center is located at Moffett Field, California, at the southern tip of San Francisco Bay, near Mountain View. The center is easily accessible from Highway 101 by taking the Moffett Blvd./NSAS Parkway exit. The center is free and open to the public on Tuesdays through Fridays from 10:00 A.M. to 4:00

> Website: www.nasa.gov/centers/ames
> Telephone: 650–604–6274

P.M. and weekends from noon until 4:00 P.M. The center is closed on Mondays and holidays. A gift shop is located near the center.

Dryden Flight Research Center at Edwards Air Force Base, Mojave Desert, California

Since the 1940s, this has been the place where highly skilled pilots test the performance of experimental aircraft, often pushing the planes to the limits of their flying capability. Nearly every U.S. military aircraft has undergone testing here. In 1947 in the skies above Edwards, Chuck Yeager broke the sound barrier in the Bell X-1, and in 1953 Scott Crossfield flew at twice the speed of sound (Mach 2) in a Douglas Skyrocket. In 1959, several test pilots from Edwards were chosen as the original *Mercury 7* astronauts. In these early days, it was not unusual for test pilots to log hundreds of flying hours per month in prototype aircraft. This resulted in an alarmingly high death rate; in fact, the base is named in honor of test pilot Glen Edwards, who died while testing the Northrup YB-49.

With the arrival of the X-15 in the 1960s, pilots set new records. In 1963, the X-15 became the first airplane to fly into space when it achieved an altitude of nearly 66 miles. In 1967, the X-15 set a speed record for

The Lunar Landing Research Vehicle (LLRV) performing a test flight at Edwards Air Force Base in 1964.

piloted atmospheric flight of Mach 6.7. Also during the 1960s, an awkward-looking machine, known officially as the Lunar Landing Research Vehicle and unofficially as the "flying bedstead," was used to study the piloting techniques necessary to land the Apollo Lunar Module on the Moon. In the 1980s, the first landings of the Space Shuttle took place here, and the site is still used as a back-up landing strip. (The last Shuttle landing here was in June 2007.) As recently as 2004, a new record was set at Edwards when the X-43A achieved a speed of 6,800 mph (Mach 9.6), the highest speed ever for an aircraft powered by an air-breathing engine.

NASA's Dryden Flight Research Center, named after Hugh L. Dryden, a prominent aeronautical engineer and deputy administrator at NASA, is actually located inside Edwards Air Force Base. The base is located next to Rogers Dry Lake Bed whose hard, flat surface provides plenty of extra landing room for the aircraft. This, along with year-round clear flying weather makes Dryden a perfect spot for testing high-performance aircraft.

Dryden has a full complement of current projects: testing the Altair, an unmanned aerial vehicle designed to carry out NASA's earth science missions; operating the ER-2 (Earth Resources), a high-altitude science aircraft; and testing the Stratospheric Observatory for Infrared Astronomy (SOFIA), an airborne observatory that makes observations in the infrared part of the spectrum. Dryden engineers and scientists are also working on various systems for the Orion Crew Exploration Vehicle.

Visiting Information

NASA Dryden sits on the western edge of the Mojave Desert, about 90 miles north of Los Angeles. Free public tours are offered every other Friday, but you must sign up for the tours in advance by calling the number below or visiting the website. The tours include a 90-minute walking tour of NASA Dryden and a 90-minute bus tour of Edwards Air Force Base. You can have lunch at the NASA cafeteria and browse through the Air Force Flight Test Center Museum and the gift shop. The tours begin at 10:00 A.M. and end at around 2:30 P.M. There is no Visitor

> Website: www.nasa.gov/centers/dryden/home/index.html
>
> Telephone: 661–277–3510

Center at Dryden; however, the Aerospace Exploration Center in nearby Palmdale is operated by Dryden and is open free of charge on Tuesdays, Wednesdays, and Thursdays from 9:00 A.M. to 3:00 P.M.

Glenn Research Center, Cleveland, Ohio

The Glenn Research Center originated in 1941 when the NACA established the Aircraft Engine Research Laboratory. After World War II the lab shifted its focus to exploring promising new propulsion technologies and changed its name to the Flight Propulsion Research Laboratory. Today, it is officially known as the John H. Glenn Research Center at Lewis Field in honor of Ohio astronaut John Glenn and George Lewis, former director of aeronautical research for the NACA. The center includes 24 major facilities and more than 500 specialized research and test labs. NASA Glenn's team of 1,300 scientists and engineers is now responsible for managing the development of

NASA

Looking upward at a Zero Gravity Facility at the Glenn Research Center. Objects can be dropped from a height of 460 feet (one and a half football fields) allowing scientists 5.18 seconds of zero gravity.

the Orion service module and spacecraft adapter. Glenn also has a major role in the Crew Exploration Vehicle and Orion environmental testing and is working on several systems for the new Ares rocket.

The NASA Glenn Visitor Center, located right inside the main gate, holds six galleries for you to explore. In one gallery, you can learn about the basic concepts of aircraft propulsion and the different kinds of engines in use today. Another exhibit highlights the Advanced Communications Technology Satellite (ACTS) that was developed here. The Solar System gallery, exploring the question "Is there life on Mars," includes a full-scale model of a Mars Sojourner Rover. In the Space Flight gallery, you can conduct microgravity experiments in a drop tower, count down to a simulated rocket launch, and try out the training module used by the heroic crew of the Space Shuttle *Columbia*. Several displays pay tribute to John Glenn and his historic Mercury flight and shuttle mission. The Visitor Center's most popular exhibit is the Apollo Command Module used on *Skylab 3*. Other Apollo displays include a model of a Saturn V rocket and a Moon rock.

Visiting Information

The Visitor Center is open every day except major holidays. Hours are Monday through Friday from 9:00 A.M. to 4:00 P.M., Saturday from 10:00 A.M. to 3:00 P.M., and Sunday from 1:00 P.M. to 5:00 P.M. There is no admission charge. The Visitor Center is open only to U.S. citizens and foreign national students in grades K–12. All adult visitors must present a government-issued photo ID. To actually tour the facilities, you'll have to visit on the first Saturday of each month from April through October. Each Saturday tour goes to a different research facility, such as the Zero Gravity Research Facility, the Icing Research Tunnel, and the Electric Propulsion Laboratory. You can make reservations for

Website: www.nasa.gov/centers/glenn
Telephone: 216-433-9653

these Saturday tours up to a month in advance by calling the number above. The center has special events on the remaining Saturdays. See the website for an event and tour schedule.

Goddard Space Flight Center, Greenbelt, Maryland

NASA's Goddard Space Flight Center, named in honor of Robert H. Goddard, the father of American rocketry, is home to the nation's largest group of scientists and engineers. They study everything here—from the Earth to the entire universe—by making observations from space. Since its beginning in 1959, Goddard has been at the cutting edge in space and Earth science research. Most of NASA's scientific research satellites and orbiting observatories, including the Hubble Space Telescope, are designed and built here. Once spacecraft are in space, Goddard operates them and monitors their orbits. One of the more recent breakthroughs made here led to a deeper understanding of how hurricanes form. Goddard also has a role in NASA's planned return to the Moon by building the Lunar Reconnaissance Orbiter.

There are no public tours of the Goddard facilities, but the Visitor Center exhibits do a good job at highlighting Goddard's contributions to our understanding of the universe. Spectacular images taken by Goddard's fleet of orbiting spacecraft make this center as much of an art gallery as a science center. The exhibits change fairly often, but one current exhibit showcases images from the Hubble Space Telescope. The images are complemented by interactive displays including a video game that assists you in calculating the distance to a galaxy and an electronic galaxy counter that helps you count the galaxies. Nearby, a camera reveals an image of your hand in the infrared.

The "Sun as Art" exhibit shows a wide variety of solar images captured by the Solar and Heliospheric Observatory (SOHO) and the Transition Region and Coronal Explorer (TRACE) missions, both of which were carried out by Goddard scientists. The images may surprise you with their colors, shapes, and beauty, enough to dispel the notion that the Sun is a boring ball of gas that never changes very much. The "Landsat-7 Earth as Art" gallery shows the Landsat satellite's images of the Earth that were selected specifically for their aesthetic appeal.

The "Science on a Sphere," a unique visual system, uses computers and video projectors to display images on a six-foot-diameter sphere suspended from the ceiling. One image shows both the day and night sides of the Earth, and, as the Earth rotates and day turns into night, splotches of light appear where the cities are. For viewing on the sphere, the center specially made the movie *Footprints,* with the theme of the human drive to explore.

Outside the Visitor Center is a small "Rocket Garden" where you can view an Apollo capsule "boilerplate" that was probably used to train astronaut crews, a collection of sounding rockets, and a Thor Delta rocket. On the first Sunday of each month, the center hosts a model rocket launch. The fenced sycamore tree is the Goddard "Moon Tree," so-called because it grew from a seed that was taken to the Moon aboard *Apollo 14.* The small garden is an "Aura Ozone Monitoring Garden," where ozone-sensitive plants develop spots on their leaves when exposed to high levels of ozone.

Visiting Information

The Goddard Visitor Center is located in Greenbelt, Maryland, a few miles northeast of Washington, D.C. In July and August, the Visitor Center is open free of charge Tuesday through Saturday from 10:00 A.M. to 5:00 P.M. The rest of the year, hours are Tuesday through Friday from 10:00 A.M. to 3:00 P.M. and Saturday and Sunday from noon until 4:00 P.M. The center is closed on all federal holidays. A gift shop

> Website: www.nasa.gov/centers/goddard/
> visitor/home/index.html
>
> Telephone: 301–286–9041

is located next to the Visitor Center. There is no food service, but picnic tables are located behind the Visitor Center.

Jet Propulsion Laboratory, Pasadena, California

Some of the most dazzling scientific discoveries of recent times have been made at the Jet Propulsion Laboratory (JPL). *Viking, Voyager, Galileo,* and *Pathfinder* are just a few of the interplanetary missions designed and built at JPL to explore the alien worlds of our solar system. These unmanned spacecraft have visited every planet and dozens of moons, and their myriad discoveries have revolutionized our understanding of the solar system. If you've watched television coverage of any of these missions, JPL's Space Flight Operations Facility (SFOF) is where you see the scientists and engineers pensively watching their computer monitors for the faraway spacecraft's faint signal, indicating that the ship has survived the journey and arrived at the destination safe and sound. If the signal comes, the control room erupts with whoops and cheers, hugs and high-fives. But if something goes wrong, the disappointment is palpable. Because of the significance of the SFOF as the hub of the communications network through which NASA controls its unmanned spacecraft in deep space, it has been designated as a National Historic Landmark.

The lab was founded in the 1930s by Caltech professor Theodore von Karman and rocket scientist Jack Parsons. Back in those days, rocket engines were often called jets or ramjets. This explains the origin of the lab's somewhat misleading name. The lab doesn't really do any research on jet engines. Today, the lab and its 5,000 employees are operated by the California Institute of Technology under a contract with NASA.

Beginning in 1964, the Space Flight Operations Facility was used for mission operations and communications with JPL's unmanned spacecraft.

Visiting Information

The absolute best time to visit JPL is during the lab's annual open house, usually held on a May weekend. If you can't make the open house, public "Visitor Day" tours are offered several times each month, usually on a Monday or a Wednesday starting at 1:00 P.M. The 2½-hour tours begin in the Von Karman Visitor Center with a 20-minute multimedia presentation that summarizes the lab's activities and accomplishments. After the presentation, you can look at the full-scale model of *Voyager* and the photographs included on the *Voyager* record. Then guides lead you into a museum where you can view models of the various spacecraft that JPL has flown or is planning to fly into space. Don't overlook the case holding the Aerogel. This material looks like a ghostly frozen mist, but it is, in fact, a solid with a density approaching that of air. Aerogel was used on the STARDUST mission to collect dust samples from a comet and interstellar dust. The tour may include visits to the Space Flight Operations Facility and the In-Situ Instruments Laboratory where the engineers put their instruments to the test. You may also get to see the Spacecraft Assembly Facility where technicians in dust-proof suits ready the robot explorers for their interplanetary journeys. You must reserve a place on the tour by calling the Public Service Office at

the number below. Call two or three
months in advance of your visit to ensure
a spot. When you arrive at the gate, just

Website: www.jpl.nasa.gov
Telephone: 818–354–9314

tell the guard you are there for a tour, and they will direct you to the visitor
parking area. You must present a photo ID. There is a small gift shop in the
building where visitors check in.

Kennedy Space Center, near Cape Canaveral, Florida

Voyages into space begin here at the John F. Kennedy Space Center (KSC),
the launch facility for all of NASA's manned spacecraft. The KSC, located
on Merritt Island near Cape Canaveral on Florida's Atlantic coast, sprawls
across more than 200 square miles and employs a total of about 15,000
people, including 2,000 NASA employees and 13,000 contract workers.
This area has been associated with rockets since 1949, when the Joint
Long Range Proving Grounds was established at Cape Canaveral to test
missiles. This new facility was deemed necessary because the White Sands
Proving Grounds in New Mexico just wasn't large enough to contain some
missile tests. In fact, in 1947, a V-2 rocket launched at White Sands went
out of control and, after a flight of only 47 miles, crashed near Juarez,
Mexico. The military decided to launch the missiles over water, but this
required tracking facilities. The first choice for a new launch facility was
the California Naval Air Station at El Centro. The plan was to launch the
missiles down the Gulf of California toward the South Pacific, but the
president of Mexico refused to grant sovereignty rights for the tracking
stations. The second choice, Cape Canaveral, was selected because the
British government was more cooperative and agreed to allow the use of
tracking stations on the Bahama Islands. Apart from the occasional hurri-
cane, the Cape had the advantage of year-round good weather and was
about as far south as you could get in the continental United States. This
proximity to the equator means that the rockets get a bigger boost from
the rotation of the Earth.

The first missile launched at the Cape in 1950 was a German V-2 rocket
with an Army WAC Corporal second stage. When President Kennedy issued
his challenge to go to the Moon, NASA needed to expand its Cape Canaveral
operations and began acquiring land on Merritt Island. After Kennedy's
assassination in 1963, the facility was renamed in his honor.

Entrance to the Rocket Garden at the Kennedy Space Center.

Today, the main launching activity at the KSC involves the Space Shuttle. The Space Shuttles are readied for launch in the giant Vehicle Assembly Building. Here, the Solid Rocket Boosters are attached to the sides of the orange External Tank, and the Shuttle itself is affixed to the External Tank. After assembly, the Shuttle is moved to the launching pad along the "crawlerway" by the massive Shuttle Crawler-Transporters, the largest tracked vehicles in the world, with a top speed of about one mile per hour. The Shuttle lifts off from one of the two launching pads at Launch Complex 39.

Visiting Information

The main attraction at the Kennedy Space Center Visitor Complex is a bus tour of the facilities. The tour, included in the price of admission, makes stops at the International Space Station Center, the Launch Complex 39 Observation Gantry, and the Apollo/Saturn V Center. Inside the International Space Station Center, you'll see the processing center where NASA is testing and piecing together the components of the International Space Station. A walk through a full-scale mock-up of a Habitation Module gives you a feel for how the crew will live, work, and sleep. Next stop is the Launch Complex 39 Observation Gantry. At the base of the gantry, you'll see a presentation detailing the steps in the launching and landing of every Space Shuttle mission. From the top of the 60-foot high gantry, you can get a panoramic view of the Vehicle Assembly Building, the crawlerway, and the two Shuttle Launching Pads, 39A and 39B. The final stop is the Apollo/Saturn V Center, a museum built around a restored Saturn V rocket, the rocket that sent the Apollo astronauts to the Moon. The bus tour is self-paced, but plan on spending a couple of hours on the tour. You can spend as much time as you want at any of the stops and then catch another bus to the next stop. The buses run about every 15 minutes. Lines for the buses can be long, so the best strategy is to arrive early and do the bus tour first.

Back at the Visitor Complex, you can visit several exhibit buildings, take a stroll through the "Rocket Garden," or take in a space movie at one of two IMAX theaters. At the "Early Space Exploration" building, you can see artifacts from the Mercury, Gemini, and Apollo programs. Visitors can glimpse into the future of space exploration, including the proposed manned mission to Mars, at the "Exploration in the New Millennium" exhibit. The "Robot Scouts" exhibit celebrates the interplanetary exploits of NASA's unmanned missions. In the Rocket Garden, Redstone, Atlas, and Titan rockets sprout skyward. These are the rockets that blasted the first NASA astronauts into space. Here, you can climb aboard the cramped quarters of the Mercury, Gemini, and Apollo capsules. At night, red-orange lights flicker at the bottom of the rockets while each fuselage is bathed in brilliant white light. Blue accents lights complete the patriotic spectrum. Free guided tours of the Rocket Garden are offered daily at 10:30 A.M. and 4:00 P.M. The Visitor Complex is also home to the Astronaut Memorial, honoring the twenty-four astronauts who made the ultimate sacrifice for space exploration. The black granite mirror makes the names of the astronauts appear to be floating in a reflection of the sky.

Two other programs are included in the Standard Admission price: the Astronaut Encounter and the Mad Mission to Mars. The Astronaut Encounter is a half hour interactive question-and-answer session with a real live astronaut. The encounters happen at various times during the day. You'll have to check the schedule at the Visitor Complex for times. If you have young children in tow, they'll no doubt enjoy the Mad Mission to Mars, a Disney-esque show where Professor Pruvitt and his friends Kelvin and WD-4D take the audience on a space voyage.

There are a couple of special interest tours offered by the KSC at an additional cost of $22. In addition to the stops on the standard KSC tour, the 90-minute NASA Up Close Tour visits the KSC industrial area and headquarters and the Shuttle Landing Facility. If you're a space history buff, then you'll probably want to take the Cape Canaveral: Then and Now Tour, where you see the launching pad and training facility used by Alan Shepard, the first American in space. The tour also stops at the Apollo Launch Pad 34, site of the *Apollo 1* fire that killed three astronauts, and at Launch Complexes 40/41 where the *Viking, Voyager,* and *Cassini* probes were launched. Both of these special tours sell out daily, so you should reserve in advance on-line or by phone.

The KSC also offers several special programs at an additional cost. The popular "Lunch with an Astronaut" program gives visitors a chance to meet an astronaut and get an autograph and a photo. The cost is $23 with a meal included. Remember, however, that you can meet an astronaut at no additional charge in the Astronaut Encounters program.

A short drive back across the bridge from the Visitor Complex is the U.S. Astronaut Hall of Fame where you can explore a collection of astronaut artifacts including Wally Schirra's *Sigma 7 Mercury* spacecraft and the *Apollo 14* Command Module.

The KSC Visitor Complex is open daily except Christmas and some launch days from 9:00 A.M. to 6:00 P.M. The Astronaut Hall of Fame is open from 9:00 A.M. to 7:00 P.M. The Standard Admission price is $31 for adults and $21 for children age 3 to 11. The Standard Admission includes the KSC bus tour and all the exhibits, shows, and movies at the Visitor Complex. Note that the admission includes as many

Website: www.kennedyspacecenter.com
Telephone for tickets and reservations:
321–449–4400

IMAX movies as you want to see. The "Maximum Access" Admission includes the U.S. Astronaut Hall of Fame for $7 extra. The KSC is a short 45-minute drive east from Orlando and is easily accessible from I-95. See the website for directions.

Stennis Space Center, Gulf Coast, Mississippi

The Stennis Space Center, named in honor of Mississippi Senator John C. Stennis, a staunch supporter of the space program, is the nation's largest rocket test facility. This is where NASA tested the rockets that sent Apollo to

NASA

The second stage of a Saturn V rocket is lifted onto the test stand at the Stennis Space Center in 1968. Other test stands can be seen in the background. The rockets were transported by barges using the waterways.

the Moon and the Space Shuttle into orbit. In fact, during the late 1960s and early 1970s, the local residents had a saying: "If you want to go to the Moon, you first have to go through Hancock County, Mississippi." In the future, the Ares rocket engines will be tested here. This site was chosen because of its unique waterway system that provides a way of transporting the huge rocket engines and loads of propellants. In addition, the site is surrounded by a sparsely inhabited area of 125,000 acres that provides a sound buffer between the local population and the thunderous roar of the rocket engines. Today, the center has evolved into a multidisciplinary facility where 30 agencies involved in space, defense, and environmental programs, have a presence.

The 25-minute bus tour takes visitors around the grounds and ends up at the StenniSphere, the visitor center for the Stennis Space Center. Inside the StenniSphere, you can take a look at actual rocket engines and scale models of various spacecraft. Some of Dr. Wernher von Braun's personal belongings are also on display here. Test your skill at landing the Space

Shuttle in a replica of the Shuttle cockpit, and take a look inside a scale model of a module of the International Space Station to see how the astronauts sleep. Stroll along "One Main Street Mars," and imagine what a future Martian settlement might look like. Live Mars-themed stage shows are performed here twice each day. Outside, you can examine a few engines that have been tested at Stennis, including a Space Shuttle main engine, a Space Shuttle solid rocket booster, a J2 engine used on the second stage of a Saturn V, and a F1 booster engine for the Saturn V's first stage. A Learjet used for NASA's remote sensing program is also on display along with a weather data buoy.

If you are very lucky, you may get to see a test firing of an engine. The public may view these test firings when they occur during the StenniSphere's hours of operation or at scheduled public test firings. Check the website or call to find out about the test firing schedule.

Visiting Information

Free shuttle bus tours of the grounds leave every 15 minutes from the Mississippi Welcome Center on I-10 at Exit 2. This exit is 45 miles east of New Orleans and 48 miles west of Biloxi. Visitors eighteen and older must

> Website: www1.ssc.nasa.gov/public/visitors/
> Telephone: 228–688–2370

present a valid photo ID. International visitors must have a passport. The StenniSphere is open Wednesday through Saturday from 10:00 A.M. to 3:00 P.M. and is closed on all major holidays. The StenniSphere includes the Odyssey Gift Shop and the Rocketeria Diner, a space-themed retro restaurant featuring local cuisine, like po-boys and gumbo. A picnic area is also available. The diner is closed on Saturdays.

Wallops Flight Facility, Wallops Island, Virginia

The Wallops Flight Facility is NASA's main facility for low-cost suborbital missions and small orbital flight projects. From here, NASA releases scientific balloons and launches small rockets called sounding rockets that carry instruments designed to take measurements and perform experiments during short flights usually lasting between five and twenty minutes. The sounding rockets operate at altitudes between the maximum altitude for balloons and the minimum altitude for orbiting satellites. The Visitor Center

NASA

A row of rocket launching pads at the Wallops Flight Facility.

has an observation deck where you can view a rocket launch. Television monitors in the Visitor Center give guests a look at launch pad activities. If the Visitor Center is closed, then visitors can view the launches from the south-facing areas on neighboring Chincoteague and Assateague Islands. Keep in mind that these rockets are small, so they may be difficult to see. Also, launch times and dates are subject to change due to weather and other factors. To get the latest launch schedule, go to the Wallops website or call the public affairs office at the number given below. The Visitor Center also has exhibits on NASA missions that explore the solar system and the universe. The Earth as Art exhibit displays beautiful images of the Earth taken by various satellites.

Visiting Information

The Visitor Center is located on Route 175 directly across from the runways. Hours are Thursday through Monday from 10:00 A.M. to 4:00 P.M. Admission is free.

Website: www.nasa.gov/centers/wallops/home/index.html
Telephone: 757–824–2298 (Visitor Center)
757–824–1579 (Public Affairs Office)

Second star to the right and straight on till morning.   Peter Pan

There are dozens of "Air and Space" museums across the country, and more are opening all the time. Most are heavy on the "Air" and light on the "Space" simply because there are a lot more aircraft and aviation-related artifacts available to museums than spacecraft and space-related artifacts. In this chapter, I have concentrated on a few museums that have the most extensive collections of space-related material or that specifically advertise themselves as space museums. The sites are listed in order according to the estimated size of the collection. A very useful website, *A Field Guide to American Spacecraft,* lists the museums and locations of every NASA spacecraft that has flown into space.

National Air and Space Museum, Smithsonian Institution, Washington, D.C.

The National Air and Space Museum (NASM) is to space what the Louvre is to art. There is simply no place on Earth with so much space and aviation history packed into a single building. Drawing an average of nine million visitors each year, the NASM is the most visited museum in the world. Upon entering the building from the National Mall, you immediately find yourself surrounded by famous airplanes and spacecraft in the museum's goose-bump inducing "Milestones of Flight" gallery. Here, you see the actual Mercury *Friendship 7* capsule that cradled John Glenn on his three orbits around the Earth; you see the actual *Gemini 4* capsule that witnessed the first spacewalk; and you see the actual *Apollo 11* command module that carried Armstrong, Aldrin, and Collins to the Moon and back. Don't miss the opportunity to reach out and touch the 4 billion-year-old Moon rock, a flat piece of black basalt brought back from the Taurus Littrow Valley by *Apollo 17*. Replicas of liquid-fueled rockets built in 1926 and 1941 by Robert Goddard himself stand in one corner. Look up to see an *Explorer 1* back-up vehicle, a replica of *Sputnik 1*, and a prototype of *Pioneer 10*. The sleek bright orange rocket plane is the actual Bell X-1 "Glamorous Glennis" that test pilot Chuck Yeager used to break the sound barrier in 1947. A Viking Mars Lander test vehicle stands on a realistic Martian surface with its robotic arm extended to take a soil sample. Suspended as if in flight is the actual Ryan NYP *Spirit of St. Louis* that Charles Lindbergh flew across the Atlantic in 1927—the first solo nonstop transatlantic airplane flight. Perhaps most significant, the *Wright Flyer* that Orville and Wilbur Wright coaxed into the air for the first powered flight at Kitty Hawk, North Carolina, in 1903 hangs from its place of honor. This is not a replica; although the canvas has been replaced, this is the actual airplane!

After ogling the national treasures in the Milestones of Flight exhibit, it's time to explore the rest of the museum. The NASM is laid out so that most of the "air" exhibits are on the west end of the museum (to your right if you entered from the National Mall), and the "space" exhibits are on the east end. Because the focus of this book is space, I describe in detail only the space-related exhibits.

Amateur astronomers will certainly enjoy the "Explore the Universe" gallery, which traces the evolution of astronomical observing techniques from early naked eye observations, to telescopes, and finally to the digital age. In the "Naked Eye" section, you encounter a collection of astrolabes (mechanical sky maps), quadrants (instruments used to measure angles in the sky), and armillary spheres (used to measure the positions of celestial objects and to teach about the celestial coordinate system). The large armillary sphere on display here is a replica of one used by the great Danish astronomer Tycho Brahe. In the late 1500s, Tycho and his assistant gathered the most accurate observational data on the positions of the planets before the advent of the telescope. Johannes Kepler later used Tycho's observational data to derive his three laws of planetary motion. In the "Telescope" section, be sure to take a look at replicas of Galileo's refracting telescope and Isaac Newton's reflector. The large 20-foot-long wooden telescope on display was built by William Herschel in 1873. One of the more unusual artifacts is a pigeon trap used by Arno Penzias and Robert Wilson to capture pigeons that had roosted in the horn of their microwave antenna. They thought that the pigeon droppings might be the source of the static they were detecting with their telescope, but, after removing the pigeons and scrubbing out the droppings, the static was still there. The bothersome static, discovered in the 1960s, turned out to be the leftover heat from the big bang. This discovery convinced astronomers that the big bang theory of the origin of the universe was correct.

The highlight of the "Digital Age" section is an actual backup mirror for the Hubble Space Telescope. This mirror, made of Corning ultra-low expansion glass, measures over eight feet in diameter and weighs 1,650 pounds. Because the mirror was never actually used, it never received the finishing touch of a reflecting aluminum coating. Other Hubble related items include the original Wide-Field/Planetary Camera (WF/PC) that was part of the Hubble when it was placed into orbit in 1990. The WF/PC is actually two cameras in one. The wide field part of the camera was used for viewing large swaths of the sky, while the planetary camera focused in on small sections of sky like those occupied by planets. Because of an optical flaw in the shape of the Hubble's main mirror, this camera was removed and replaced with the WF/PC-2, which optically compensated for the flaw and gave the telescope its 20/20 vision. The stunning photographs from the Hubble Space Telescope were taken with the WF/PC-2.

The "Exploring the Planets" gallery takes visitors on a tour of the solar system. One exhibit shows the relative size of the planets, thirteen of the largest moons, and an arc representing the sphere of the Sun. Notice the puniness of the planets compared to the enormity of the Sun. In fact, more than one million Earths could fit into the volume of the Sun! Each of the four inner earthlike planets is featured in its own exhibit, rich with detailed information and photographs. Of particular interest is the Mars exhibit where you can see a globe of Mars constructed by Percival Lowell who mistakenly thought he saw canals on Mars. You'll also see replicas of instruments used on the Mars Viking Landers, including a camera, a weather sensor, and a gas chromatograph mass spectrometer used to determine the chemical composition of the atmosphere and search for organic compounds. A full-scale replica of the *Voyager* spacecraft hovers over the Outer

Planets exhibit, which highlights some discoveries made by *Voyagers I* and *II* and other spacecraft that have visited these gaseous giants. Don't miss the duplicate of the *Voyager* record containing the sounds of Earth. Pluto's demotion to a "dwarf planet" is noted along with a replica of the photographic plate with which Clyde Tombaugh discovered Pluto. The "What's New" exhibit reports on the latest solar system discoveries by NASA's unmanned spacecraft.

The "Apollo to the Moon" gallery holds nearly 200 artifacts from the Apollo and related NASA missions including food, tools, instruments, personal items, and clothing. Three of the most significant items are the Lunar Roving Vehicle (LRV), the spacesuits worn by Armstrong and Aldrin during the first lunar landing, and the Saturn V F-1 engine. The LRV was used on *Apollo 15, 16,* and *17* to give the astronauts greater freedom to explore more of the lunar surface. On the three Apollo missions, the LRVs were driven more than 50 miles. The LRVs were powered by two 36-volt silver-zinc batteries that could operate the vehicles for up to 78 hours. The LRVs had a maximum speed of nine miles an hour and a maximum range of about 57 miles. The wheels of the LRVs could be folded under the chassis and were stored in the descent stage of the lunar module. The LRV on display here is one of eight test vehicles built for NASA by the Boeing Company. Standing near the LRV are the actual space suits, helmets, and gloves that Armstrong and Aldrin wore on the first Moon landing on July 20, 1969. The overshoes and portable life support systems were left on the Moon to reduce weight. The suits had to be flexible enough to allow the astronauts to walk and to pick themselves up if they fell and yet durable enough to resist ripping or tearing. The suits have 25 layers of materials to protect the astronauts from the harsh lunar environment, including micrometeoroids and temperatures ranging from -250 °F to 230 °F. The outermost layer of the suit is made from a white Teflon coated fiberglass called beta cloth. The suits weighed a bulky 180 pounds on Earth but a manageable 30 pounds on the Moon; each suit was individually tailored to fit the astronauts.

At the end of your walk through this gallery, you see an F-1 rocket engine used to propel the Apollo Saturn-V rocket into space. A clever arrangement of mirrors is used to multiply the single engine into a cluster of five engines as they would have appeared at the base of the rocket. Each F-1 engine delivered more than 1.5 million pounds of thrust and burned more than 100,000 gallons of liquid oxygen and kerosene fuel in the short 2.5–minute lifetime of the Saturn-V first stage. As you exit the gallery, take

Ritu Manoj Jethani/Shutterstock

An Apollo Lunar Module on display at the National Air and Space Museum.

a moment to watch the video of President John Kennedy's congressional address in which he challenges the nation to land a man on the Moon and return him safely back to Earth.

At the east end of the building sits an actual Lunar Module (LM), one of twelve built for NASA by the Grumman Aerospace Corporation. This particular LM had originally been scheduled to go into orbit to test rendezvous, docking, and separation procedures, but a previous successful test made a second test unnecessary. The LM consists of two stages: the silver and black ascent stage, containing the crew, sits atop the shiny gold descent stage. A rocket on the descent stage ensured a gentle lunar landing. After completing their explorations, the astronauts would climb back into the ascent stage and blast off from the descent stage leaving it behind on the Moon. After docking with the command module, the ascent stage was jettisoned and eventually crashed into the Moon.

Suspended above the Lunar Module are representatives of three unmanned lunar exploration programs that blazed the trail for Apollo. The Ranger series of lunar probes sent back the first close-up pictures of the Moon's surface before crashing. This is a replica of Ranger 7, the first successful probe in the Ranger program (Rangers 1 through 6 failed). Next came the Surveyor program. From 1966 through 1968, five successful Surveyors tested soft-landing techniques, sent back nearly 90,000 pictures of the lunar surface, and chemically analyzed the lunar soil. The Surveyor on display here was used for ground tests. Finally, the Lunar Orbiter series photographed 95 percent of the Moon's surface helping NASA scientist identify landing sites for the Apollo missions. The Lunar Orbiter on display here was also used for ground tests. Clementine, the other lunar probe on display in this area, didn't go to the Moon until 1994. Clementine spent two months in lunar orbit mapping the surface and helping answer some lingering questions about the Moon. Data collected by Clementine first hinted at the possibility of water ice on the Moon, a discovery that was later confirmed by the Lunar Prospector. After visiting the Moon, Clementine was supposed to fly by an asteroid and determine its mineral content. The name "Clementine" comes from the song "My Darlin' Clementine" about a miner's daughter. After its encounter with the asteroid, the probe would be "lost and gone forever"; the fly-by, however, failed due to a malfunction.

The theme of the Space Race Gallery is the contest between the United States and the former Soviet Union for dominance in space. The rockets and missiles on display here, relics of this race, are the German V-2, Vanguard, and Jupiter C rockets and Minuteman III and Tomahawk cruise missiles. The centerpiece of the gallery is the Apollo-Soyuz spacecraft. After more than a decade of an intense space rivalry, the United States and the USSR, in a gesture of good will, agreed in 1972 to embark on a cooperative mission in space called Apollo-Soyuz. ("Soyuz" is the Russian word for "union.") The agreement was fulfilled in 1975 when an *Apollo* spacecraft rendezvoused in orbit with a Russian *Soyuz* craft. Using a docking module collaboratively designed by American and Russian engineers, the two spacecraft linked together, and *Apollo* commander Thomas Stafford floated through the hatch and shook hands with Soviet commander Aleksei Leonov. For nearly two days, the crews visited each other's spacecraft, shared meals, and conducted joint scientific experiments. The Apollo-Soyuz mission, marking the last flight of the *Apollo* spacecraft, was the last manned NASA mission until the first space shuttle launch in 1981.

Another spacecraft prominently displayed in this gallery is a Skylab back-up ship. Launched in 1973, Skylab was constructed from the third stage of a Saturn V launch vehicle. Each of the three missions carried a crew of three astronauts who collected scientific data and proved that humans could live and function in space for months at a time. You can walk through this Skylab and see the crew compartments where the astronauts ate, slept, and took showers.

The gallery also features various space clothing, including a pressure suit worn by Yuri Gagarin, the first person in space, and flight suits worn by John Glenn, Guion Bluford, the first African American in space, and Sally Ride, the first American woman in space.

Finally, don't miss the Hubble Space Telescope Dynamic Test Vehicle. This is a full-scale mock-up of the spacecraft built by the Lockheed Corporation, the company that eventually won the contract to build the orbiting observatory. Initially used for a variety of feasibility studies and later for thermal and vibrational tests, the test vehicle eventually evolved into a simulator for developing maintenance and repair procedures to be performed by astronauts.

The Rocketry and Space Flight gallery traces the history of rockets from their origins in thirteenth-century China until today. One of the most unusual objects on display is a model of a very early rocket-like device that was described by a Syrian scholar in 1280 as a self-propelled combusting egg. Jumping ahead to the early 1800s, William Congreve in England designed rockets that were effectively used as weapons. Congreve rockets like the one on display were used during the battle of Fort McHenry in 1814. Francis Scott Key watched the battle, saw the rockets, and later made reference to "the rocket's red glare" in the "Star-Spangled Banner." In the 1840s, Englishman William Hale improved the stability and accuracy of rockets by engineering them to rotate as they flew. Hale rockets of the type on display were used by the British during the Crimean War and by both the North and the South in the American Civil War. A rocket built by Robert Goddard in May 1926 can be viewed along with a 1928 Goddard "hoopskirt" rocket. There's an exhibit on rockets in science fiction that includes some Buck Rogers toys and a model of the spacecraft described by Jules Verne in his book *From Earth to the Moon*. Scientists have often been inspired by reading science fiction as children. In fact, Robert Goddard's scientific imagination was stirred by reading H. G. Wells's *War of the Worlds*.

A number of rocket engines are exhibited, including a cross-section of an RL-10 rocket engine that provides a view of the internal structure of the engine. The most exotic engine on exhibit is surely the Project Orion Test Vehicle Rocket Engine. Project Orion was a serious attempt during the late 1950s and early 1960s to design a nuclear-powered spaceship. The basic idea involved detonating small nuclear bombs one after another against a large steel plate attached to the ship by giant shock absorbers. (No, I'm not making this up.) The explosions would transfer momentum to the ship and allow it to achieve extremely high speeds. Such a ship would be able to make a round trip to Mars in about four weeks, compared to a full year for NASA's current chemically powered rockets. Scientists were seriously developing Project Orion, which appeared entirely practical, until it was shut down in 1965 due to the Partial Test Ban Treaty that prohibited nuclear explosions in space.

The exhibit ends with an impressive collection of space suits that trace the evolution of the pressurized space suit from early deep sea diver's suits through spacesuits used in Mercury, Gemini, and Apollo. The exhibit courageously addresses the indelicate question often asked by school children: How do the astronauts go to the bathroom? A "Urine Collection and Transfer Assembly" is on display here as is a "Fecal Containment System" so both Number 1 and Number 2 are covered!

The "Looking at Earth" gallery explores the practical uses of looking at Earth from the vantage point of space. Orbiting satellites, our "eyes in the skies," help us forecast the weather, monitor crops, locate resources, map the terrain, and gather intelligence. Several satellites on display in this gallery include a TIROS II (Television and Infrared Observation Satellite) test vehicle used for meteorology, an ITOS (Improved TIROS Operational System) test vehicle that surpassed the original TIROS system by storing and directly transmitting television and infrared images, and a half-scale GOES (Geostationary Operational Environmental Satellite) model used to closely monitor weather patterns above a particular point on Earth's surface. Also on display are scanning and mapping instruments used on the LANDSAT satellite. You can check out satellite images of your own state on the interactive video screens. Probably the most famous object on display in this gallery is a Lockheed U-2C spy plane, one in a series of spy planes used during the cold war for high altitude photo-reconnaissance. In 1960, a U-2 piloted by Francis Gary Powers was shot down over the Soviet Union. The U-2 also kept watch on the Soviet missile build-up that led to the Cuban missile crisis in 1962, monitored nuclear tests in China, and collected intelligence over Vietnam and the Middle East.

Finally, "Star Trek" fans will want to pay homage to the original Starship *Enterprise* model on display in the museum gift shop. This 11-foot-long, 200-pound model, made mostly of poplar wood and plastic, was used in the filming of the original "Star Trek" television series, which ran from 1966 to 1969.

Visiting Information

The National Air and Space Museum is part of the Smithsonian Institution and is located on the National Mall in Washington, D.C. The street address is Independence Avenue at 4th Street, SW. The NASM is open from 10:00 A.M. to 5:30 P.M. every day except Christmas. Admission is free. The closest Metro stops are Smithsonian and L'Enfant Plaza. Free docent-led tours are offered daily at 10:30 A.M. and 1:00 P.M. The tours leave from the Welcome Center and last about an hour. The Lockheed Martin IMAX Theater offers several different movies daily. IMAX tickets are $8.50 and can be purchased in advance online. Tickets for shows at the Albert Einstein Planetarium are also $8.50 and available for purchase online. The planetarium does offer a few free shows every week. Flight simulator rides are available for $7.00. Food is available in the glass enclosed Wright Place Food Court located at the east (space) end of the building. The food court

> Website: www.nasm.si.edu
> Telephone: 202–633–1000

offers menu items from McDonald's, Boston Market, and Donato's Pizza. Your credit card could get a real workout at the three-level Museum Store where you can buy all sorts of space- and aviation-related items.

National Air and Space Museum's Udvar-Hazy Center, Virginia

As big as the National Air and Space Museum's building on the National Mall is, it can only display about 10 percent of the museum's extensive collection of aircraft and space artifacts. Another 10 percent of the collection is on loan, but that still leaves 80 percent of the collection. The Udvar-Hazy Center was built so that the complete collection could be put on display. The center will eventually include a restoration area where the public can watch specialists restore historic aircraft and artifacts. The center is named after Steven F. Udvar-Hazy, chairman and CEO of a company that owns and leases commercial aircraft around the world. Udvar-Hazy donated $65 million to build this new center, which opened in December 2003 and has been drawing more than a million visitors each year. In 2008, the center had 141 aircraft, 148 large space artifacts, and more than 1,500 smaller objects on display. Most of the center's display space is devoted to aviation and visitors

The Space Shuttle *Enterprise* on display at the Udvar-Hazy Center.

should definitely not miss seeing the *Enola Gay,* the plane that dropped the atomic bomb on Hiroshima. Again, since the focus of this book is space, the detailed descriptions will be limited to the space-related exhibits on display in the James S. McDonnell Space Hangar.

The centerpiece of the space exhibits is the Space Shuttle *Enterprise.* This was the very first Space Shuttle, designed for atmospheric tests rather than space flight; therefore, it was built without engines or a functional heat shield. The planned name for this shuttle was "Constitution," but a write-in campaign by fans of the "Star Trek" television series convinced NASA officials to change the name to "Enterprise." When the newly completed *Enterprise* was rolled out of its hangar in 1976, most of the cast of the original series greeted it, and the "Star Trek" theme music was played.

The *Enterprise* was ferried to the Dryden Flight Research Center at Edwards Air Force Base in California to begin a series of ground and test flights to see how the shuttle handled on approach and landing. For most of the tests, the *Enterprise* rode piggyback on top of a Boeing 747, but for the last few, it glided on its own with an astronaut crew at the controls. These last tests verified the flight characteristics of the Shuttle itself under conditions simulating a return from space. Following these tests, the *Enter-*

prise was transported to the Marshall Space Flight Center in Alabama for vibration testing and then on to the Kennedy Space Center where it was mated with an external fuel tank and solid rocket boosters and tested in a launch configuration.

After all the critical testing was done, the *Enterprise* was partially disassembled so that some parts could be used in other shuttles. Then, the *Enterprise* embarked on a European tour with stops in England, France, Italy, and Germany. Following appearances in Canada and a few of the states, the *Enterprise* was turned over to the Smithsonian in 1985. After the breakup of the space shuttle *Columbia* upon reentry in 2003, the investigation team removed a panel from the wing of the *Enterprise* for testing. The test involved shooting a piece of foam at the panel to see if any damage would be done. The impact permanently deformed a seal. This and other tests led investigators to conclude that the breakup of the *Columbia* had been caused by a piece of foam that broke off from the external fuel tank during launch and hit the leading edge of the wing.

Sitting near the *Enterprise* is a shiny, silver Airstream trailer that is, in fact, a Mobile Quarantine Facility (MQF). NASA built four of these specially outfitted trailers to hold the Apollo astronauts in isolation after returning from the Moon, just to make sure the astronauts had not been infected with any Moon germs. This particular MQF was used by the crew of *Apollo 11* after their historic lunar landing. They were held inside for 65 hours while traveling from the aircraft carrier to the Lunar Receiving Laboratory at the Johnson Space Center in Houston.

Check out the Goddard 1935 A-series rocket on display. Goddard's early rockets had not been able to achieve the altitude or maintain the stability that he and his supporters had hoped for. The A-series rocket was more promising—a high-altitude rocket that could maintain a steady, vertical flight, reaching altitudes that exceeded a mile and registering speeds that exceeded 700 mph. Goddard tested his rockets at his laboratory near Roswell, New Mexico. After a series of successful launches, Goddard pieced together a complete A-series rocket and donated it to the Smithsonian. Other space artifacts include a Mars Pathfinder Lander Prototype and Airbags, a Mars Sojourner Rover Model, and a model of the mother ship from the movie *Close Encounters of the Third Kind*.

For astronomy buffs, several early space telescopes are on display, including the balloon-born Stratoscope and the Copernicus Orbiting Telescope. The ground-based Caltech 2.2 micron telescope was used to map the sky in the infrared region of the spectrum. Expecting to find a few hundred infrared sources, astronomers were astonished to find tens of thousands. Of particular interest is the Ritchey Mirror Grinding Machine, a device built by George Ritchey at the Yerkes Observatory in the late 1890s. This machine has been used to grind mirrors for numerous telescopes for more than a century. George Hale used it to grind a 60-inch mirror for the Mount Wilson observatory. In the 1970s, it was used to grind the 40-inch mirror for the Nickel telescope at Lick Observatory. Most recently, it was used to make a variety of optical elements for the Keck Observatory in Hawaii.

Visiting Information

The Udvar-Hazy Center is located next to Washington Dulles International Airport in Chantilly, Virginia. No Metro rail stops or Metro bus stops are nearby. A Virginia Region Transportation Association (VRTA) shuttle bus stops at the center. The Udvar-Hazy Center is 28 miles from the Air and Space Museum on the National Mall. If you are at the National Mall and want to get to the Udvar-Hazy Center and you don't have a car, the VRTA bus may be an option. Check with the information desk at the museum, and the guides can tell you which bus to catch and where. The bus trip takes about 90 minutes. Another option is an expensive cab ride. There is no charge for entering the museum; however, there is a $12 parking fee. Museum hours are 10:00 A.M. to 5:30 P.M. daily. The museum is closed on Christmas Day. Facilities include a museum store, simulator rides, an IMAX Theater, and a dining hall.

> Website: www.nasm.edu
> Telephone: 202–633–1000

Kansas Cosmosphere and Space Center, Hutchison, Kansas

It may surprise you (as it did me) to learn that one of the very best space museums in the country is located way out on the plains of Kansas in the city of Hutchison. The Kansas Cosmosphere and Space Center is home to a space artifact collection that is second only to the National Air and Space Museum in Washington, D.C. It is one of only three places in the world where you can see a complete set of Mercury, Gemini, and Apollo spacecraft that have actually traveled into space. This impressive trio includes the Mercury "Liberty Bell" 7, Gemini X, and the famous Apollo 13 "Odyssey" command module. The museum boasts the largest collection of Russian space artifacts that can be seen outside of Russia, including a flown Vostok space-craft. Pretty impressive, huh?

So how did a space museum wind up in Kansas? The origin of the Cosmosphere can be traced to 1962, when Patricia Carey, a local resident and lover of science, motivated by the launch of *Sputnik,* bought a used star projector and dome and, with a staff of high school volunteers, founded the Hutchison Planetarium in the poultry building on the Kansas State Fairgrounds. Four years later, the planetarium moved into the new science building on the campus of what is now Hutchison Community College. In the 1970s, the planetarium board asked Max Ary, a space expert who had worked at the planetarium while attending college, if he had any ideas for a possible museum. As luck would have it, Ary was serving on a Smithsonian committee working to find a repository for tens of thousands of space artifacts released after the end of the Apollo program. With Ary as its director, the Cosmosphere was born in 1980; what had begun as a local planetarium, Ary transformed into a world-class space museum. Sadly, in 2005, Ary was charged and later convicted of stealing artifacts from the collection and selling them for personal profit.

Today, the Cosmosphere is home to SpaceWorks, the only permanent facility in the world that specializes in restoring space artifacts. In the early days, the Cosmosphere could not compete with larger museums with deeper pockets in acquiring space artifacts that were in prime condition. Thus, the Cosmosphere purchased cheaper items in disrepair and completed their own restoration. The staff sought out original manufacturers' plans and blueprints and developed great skill at reconstructing space artifacts. Now, SpaceWorks services not only the Cosmosphere but also museums, science centers, and even the movies. Director Ron Howard hired the company to design and build the spacecraft interiors for the movie *Apollo 13.* According to Howard, "I could have used NASA to reconstruct the interiors, but the folks at SpaceWorks could make it look more authentic." SpaceWorks worked on the *Apollo 13* Command Module Odyssey and on Gus Grissom's *Liberty Bell 7* capsule after it had been sitting on the bottom of the Atlantic Ocean for more than 30 years. Visitors can see part of the SpaceWorks facility as they tour the museum.

Upon entering the museum's lobby, you find yourself in the dark shadow of a flown SR-71 *Blackbird,* a famous spy plane used to keep a watchful eye on the Soviets during the Cold War. Suspended from the ceiling next to the *Blackbird* is a Northrup T-38 Talon, a supersonic training jet used by

NASA pilot astronauts to maintain their flying skills and by mission specialists to get familiar with high performance jets. Also on display is a full-scale mock-up of the left side of the Space Shuttle.

The Cosmosphere's Museum of Space consists of four major galleries: the German Gallery, the Cold War Gallery, the Early Spaceflight Gallery, and the Apollo Gallery. The galleries are arranged chronologically to trace the history of space flight and the museum attempts to place every artifact into some kind of historical and sociological context.

The German Gallery describes the beginnings of rocketry with an emphasis on the German rocket program during World War II under the direction of Dr. Wernher von Braun. This is the only place in the world where you can see authentic restored versions of both the V-I "buzz bombs" and a V-2 rocket. A Messerschmidt Me-163 Komet engine, the world's first mass-produced rocket engine, is also on display.

The Cold War gallery chronicles the initial stages of the rivalry between the U.S. and Soviet space programs, including the Cold War stalemate and the early days of the space race. Exhibits include a Sonic Wind Rocket Sled, a Redstone Atomic Warhead that was found rotting away in an Alabama warehouse, replicas of the Russian Sputnik I and II satellites, and a section of the wall that separated the East German town of Boeckwitz from the West German town of Zicherie. Don't miss the actual engine from the Bell X-1 Glamorous Glennis that Chuck Yeager used to break the sound barrier along with the replica of the Bell X-1 that was used in the movie *The Right Stuff*.

The story of space exploration continues in the Early Spacecraft Gallery, which corresponds to the period from the early to mid-1960s. Here you can climb the gantry to get a look at a replica of the 109-foot-tall Titan II rocket. Audio aids allow you to hear the chatter from Mission Control and the roar of the rocket's engines. A replica of the 83-foot-tall Redstone rocket used for the Mercury program is also on display. You can view an actual unmanned Russian Vostok (Russian for "east") that was flown in space, the type of spacecraft that took Yuri Gagarin into space, as well as a full-scale engineering model of the Voskhod (Russian for "rising"). There were only two Voskod missions, the main purpose of which was to achieve more Soviet "firsts" in space—most notably, the first multiperson crew and the first spacewalk. One of the most prized exhibits is the spacesuit worn by Svetlana Savitskaya, the first woman to walk in space.

Two of the three jewels of the Cosmosphere's collection, the Gemini 10 capsule and the Mercury "Liberty Bell" 7 capsule are on display in the Early

Spacecraft Gallery. The Gemini X mission was flown in July 1966 by John Young and Michael Collins (who later flew on the historic *Apollo 11* Moon mission). The Gemini X completed 43 orbits while the astronauts practiced rendezvous and docking procedures, performed two spacewalks, and completed about a dozen experiments.

The Mercury "Liberty Bell" 7 (the name comes from the bell-like shape of the capsule and the "7" is in recognition of the teamwork of the seven original Mercury astronauts) was flown by astronaut Gus Grissom on July 21, 1961. This was the second U.S. manned spaceflight; a brief 15-minute suborbital flight covering a distance of about 300 miles. The capsule featured several design changes from the original Mercury spacecraft including a lightweight explosive side hatch, added at the request of the astronauts who complained that the original method of exiting through the top of the capsule was too difficult. The explosive hatch could be activated by the astronauts by pressing a plunger that was positioned about six to eight inches from the astronaut's right arm. The plunger had a pin inserted in it to prevent accidental release; with the pin in place, it took about 40 pounds of force to activate the hatch, but without the pin, only about five to six pounds of force was required. From the outside, a rescuer could release the hatch simply by pulling on a lanyard.

The flight of the Liberty Bell 7 went pretty much according to plan until the splashdown in the Atlantic Ocean near the Bahama Islands. Upon impact, Grissom took off his helmet and told the pilot of the recovery helicopter to give him a few minutes to record his cockpit panel data. After logging the data, Grissom radioed the helicopter and gave the go-ahead for pickup. Grissom removed the pin from the explosive hatch plunger, laid back in his seat, and waited. According to Grissom, "I was lying there, minding my own business, when I heard a dull thud." The explosive hatch had somehow blown open and sea water was sloshing into the capsule. Grissom later insisted that he had not touched the hatch plunger. He grabbed the instrument panel and lunged out of the sinking capsule into the ocean.

The recovery helicopter's crew, focusing their attention on recovering the capsule, left Grissom to fend for himself. This action was not the result of some cold-hearted military mindset, but rather of training. During practice exercises, the helicopter pilots had noticed that the astronauts seemed perfectly comfortable and content in the water. The helicopter personnel attached a cable to the recovery loop on the top of the capsule and began trying to lift the capsule out of the water. At this point the wheels of the copter were already in the water. When Grissom observed that the helicopter was having some difficulty, he swam back toward the capsule to see if he could help. Once he saw that the cable was properly connected, he knew there was nothing he could do. As the helicopter strained to lift the water-filled capsule out of the water, a warning light alerted the crew that the engines were on the verge of failure. The pilot had no choice but to release the cable and raise the personnel hoist that had been lowered for Grissom. The helicopter flew away, and the task of rescuing Grissom fell to a back-up helicopter. Meanwhile, Grissom's spacesuit was losing air, and, as a result, he was losing buoyancy. He was having trouble treading water and was periodically bobbing under the swells. The waves generated from the helicopter blades were making a bad situation even worse. Finally, the second helicopter dropped a "horse-collar" lifeline to Grissom. Exhausted, he flopped into the sling backward and was lifted awkwardly to the safety of the copter as his Liberty Bell capsule sank to the bottom of the ocean. An officer on board the aircraft carrier suggested that a marker could be placed at the spot where the capsule had sunk so that it might be recovered later, but his suggestion was rejected by the admiral who thought that ocean was too deep at that location for recovery to be practical.

The mystery of the blown hatch haunted Grissom for the rest of his career. To add insult to injury, Grissom's wife Betty was not invited to the White House after his mission as Alan Shepard's wife had been following the first Mercury flight. While Grissom steadfastly maintained that he did not touch the plunger, others suggested he had panicked and prematurely hit the plunger. A review board cleared Grissom of any responsibility regarding the hatch, but questions about how the hatch was blown lingered. Several years after the flight, Grissom suggested that the hatch may have blown if the external lanyard had come loose. Grissom's innocence is supported by the fact that on all the other missions where the hatch was blown manually, the astronaut sustained a slight bruise or cut on the hand, whereas Grissom did not.

Gus Grissom died in the tragic 1967 Apollo 1 launch pad fire, along with astronauts Ed White and Roger Chaffee. Ironically, the engineers who designed the Apollo spacecraft had rejected the use of an explosive hatch because they had discovered that explosive hatches could, in fact, fire without being triggered. The inability to open the hatch quickly contributed to the death of the Apollo 1 astronauts. Subsequent Apollo capsules were redesigned so that the hatch could be hastily opened.

The Liberty Bell sat on the ocean floor three miles below the surface until, after a 14-year search, it was recovered on July 20, 1999, just a day before the thirty-eighth anniversary of the flight. The expedition was led by underwater salvager Curt Newport, who had also worked on recovery efforts for TWA flight 800, the Space Shuttle *Challenger,* and, most famously, the *Titanic.* The search effort was financed by the Discovery Channel which also produced a documentary film about the effort. The capsule was inspected by NASA, cleaned, restored, and put on permanent display here at the Cosmosphere. It was estimated that the infamous hatch should have come to rest within a mile of the capsule, but the expedition decided it would be too hard to find, so the hatch was never recovered. The whole idea of recovering the Liberty Bell went against the wishes of Grissom's widow, who said, "I know Gus didn't do anything wrong, but even if they do find it, it won't prove a thing."

The Apollo Gallery tells about mankind's greatest journey, the Apollo missions to the Moon. Some of the artifacts on display here are a scale model of a Saturn V rocket, a lunar module, a replica of a lunar rover, flown Apollo spacesuits, and a Moon rock collected by the Apollo 11 astronauts. Also on display is an Apollo "White Room" that was attached to the capsule and served as a waiting area for the astronauts before entering the spacecraft before launch.

The crown jewel in the collection at the Kansas Cosmosphere has to be the famous Apollo 13 command module "Odyssey" located here in the Apollo Gallery. The Odyssey is displayed so that visitors can walk around it, peek inside it, and examine the damage done to the heat shield. The voyage of Apollo 13 was immortalized in the Academy Award–winning movie *Apollo 13* starring Tom Hanks. If you somehow missed the movie, stop reading, run down to your local video store, and rent it. The Cosmosphere built some of the props for the movie and the film is shown continuously as part of the exhibit.

Apollo 13, the third manned mission to attempt a landing on the Moon, was launched on April 11, 1970. The three man crew consisted of Commander James A. Lovell, John Swigert, and Fred Haise. Their destination was the Fra Mauro Highlands on the Moon. Fifty-six hours into the flight, at a distance of about 200,000 miles from Earth, Mission Control requested that the crew perform a "cryo-stir," a procedure that would stir the oxygen tanks to prevent the oxygen slush from stratifying. Some damaged wires powering the stirrer created a short circuit, and the insulation caught fire. The fire heated the surrounding oxygen, which, in turn, increased the pressure, causing the #2 oxygen tank to explode. The explosion blew off an entire section of the exterior wall of the Service Module and damaged the #1 oxygen tank, causing it to rapidly release precious oxygen into the vacuum of space. Swigert alerted Mission Control: "Okay, Houston, we've had a problem here." Mission Control replied: "This is Houston. Say again please." Then Commander Lovell uttered the now famous words: "Houston, we've had a problem . . ."

Within three hours of the explosion, all the oxygen in the tanks had been vented out into space. Because both the Command and Service Modules relied on the oxygen to generate electricity, very little electrical power was available for the spacecraft. The Command Module did have batteries that would last for ten hours, but the module needed that power for reentry. To preserve the needed electricity, the crew "powered-down" the Command Module and climbed into the Lunar Module, which would serve as their lifeboat. Landing on the Moon was now out of the question. The crew and the engineers at Mission Control turned their attention to returning the crew safely back to Earth.

To get back as quickly as possible, the spacecraft would make a single pass around the Moon allowing the Moon's gravity to slingshot the ship back toward the Earth on a "free return" trajectory. Maneuvering the spacecraft into the right position for this swing around the backside of the Moon required firing the rocket engines. But because the extent of the damage to the service module was unknown, firing the main rockets engines was deemed too risky. Instead, the engineers planned a course correction by using the lunar module's descent rocket engine. Once the spacecraft emerged from behind the Moon, the engine was fired again to accelerate the ship back toward the Earth. As the ship approached the Earth, the rocket was fired briefly for a final time to make a minor course adjustment.

The crew's lifeboat, the Lunar Module, was designed to support two men for two days. Now, it had to keep three men alive for four days. One major problem was related to the fact that humans inhale oxygen and exhale carbon dioxide (CO_2). In an enclosed space, the amount of CO_2 would increase to a dangerous level. To prevent this, the Apollo Spacecraft was equipped with CO_2 scrubbers that would remove the CO_2 from the air inside the spacecraft. The scrubbers used a chemical called lithium hydroxide, but the Lunar Module's supply of lithium hydroxide would not last four days. Additional canisters of lithium hydroxide were onboard the command module, but the canisters were the wrong shape to fit into the lunar module's receptacle. The engineers had to figure out how to fabricate an adapter from the materials available in the spacecraft. Yet another concern was water condensation in the temporarily vacated Command Module due to the colder than expected temperatures resulting from the power being turned off. The water might damage the electronics, but this would only become known when the power to the Command Module was switched back on.

As the ship approached the Earth, the crew prepared for reentry by abandoning the Lunar Module that had served them so well and crawled back into the Command Module. They successfully powered-up the Odyssey and jettisoned the Service Module and the Lunar Module. As it slowly floated away, the astronauts could see for the first time the extent of the damage to the Service Module. Photographs of the Service Module were taken for later analysis.

The world breathed a collective sigh of relief when the crew safely splashed down in the ocean. Fred Haise had to be treated for a urinary tract infection caused by a lack of water and the difficulty of disposing of urine. The crew had to store waste onboard rather than ejecting it out into space because the release of the waste could have affected the flight path, which would have required another course correction.

Superstitious people have made much of the fact that this was Apollo 13, and the number 13 is supposedly unlucky. But the safe return of the astronauts was exceedingly lucky in a number of ways. For example, if the explosion had occurred on the return trip from the Moon and after the Lunar Module had been jettisoned, the crew could not have used it as their lifeboat and their chances of survival would have been slim. In fact, NASA likes to refer to Apollo 13 as a "successful failure." The safe return of the Apollo 13 astronauts is testimony to the ingenuity and problem-solving ability of the NASA engineers and flight controllers and to the ability of the astronauts to function in the face of death.

Artistically minded scientific travelers will want to admire the stained glass creation gracing the rotunda of the Cosmosphere. At the center of the work stands an astronaut on the Moon with outstretched arms and legs, a pose reminiscent of Leonardo da Vinci's Vitruvian Man. The astronaut is surrounded by images of a galaxy, a star, a planet, a Saturn rocket, and the Space Shuttle. In the background are a shining Sun and a waving American flag. Directly above and beneath the astronaut are scrolls with the Latin phrase "Ad Astra per Aspera" and its translation "To the Stars through Difficulties." The phrase just happens to be the Kansas state motto and, coincidentally, NASA's motto. The work pays tribute to the seventeen American astronauts whose lives have been lost in the exploration of space. Incorporated into the work at the bottom of the glass is a wiring block from the Apollo 1 pad and flown shuttle tiles from the Space Shuttles Challenger and Columbia. The names of the fallen astronauts are etched into the top of the glass.

In addition to the Museum of Space, the Cosmosphere is home to an IMAX Theater and the state-of-the-art Justice Planetarium. The museum also has a science demonstration show called Dr. Goddard's Lab. The show is presented once every day in a theater-like setting representing rocket pioneer Robert Goddard's 1930s-era lab. The demonstrations are related to Goddard's experiments with liquid fueled rockets.

Visiting Information

The Kansas Cosmosphere and Space Center is located in Hutchison, Kansas, about an hour's drive west of Wichita at 1100 North Plum Street in the middle of a quiet, residential neighborhood. The Cosmosphere is open every day except Christmas. Hours are Monday through Thursday from 9:00 A.M. to 6:00 P.M., Friday and Saturday from 9:00 A.M. to 9:00 P.M., and Sunday from noon until 6:00 P.M. Single Venue tickets good for the Space Museum, IMAX, or planetarium, are $8 for adults and $7.50 for seniors 60 and over and for children ages 5 to 12; children 4 and under are free. All inclusive "Mission Passes" are $13 for adults, $12 for seniors, and $10.50 for children. The Lunar Outpost food court offers sandwiches, Papa John's Pizza, and Space Dots Ice Cream. Space-related souvenir items are available at the Cargo Bay Store.

> Website: www.cosmo.org
> Telephone: 1–800–397–0330 or
>   1–620–662–2305

U.S. Space and Rocket Center at the Marshall Space Flight Center, Huntsville, Alabama

The George C. Marshall Space Flight Center (MSFC) is NASA's main facility for the development of space propulsion systems. This is where NASA builds its rockets. Established in 1960, the center's first director was none other

Astronauts practice construction techniques needed to build a space station in the Neutral Buoyancy Simulator.

than German rocket genius Werner von Braun. The Redstone rockets that lifted the first Mercury astronauts into space and the Saturn series rockets that sent the Apollo astronauts to the Moon were developed here at the MSFC. More recently, the center was responsible for designing the propulsion systems that power the Space Shuttle, including the external fuel tank, the twin solid rocket boosters, and the orbiter's triad of engines. The MSFC has managed other notable NASA projects, including the Lunar Roving Vehicle and the Hubble Space Telescope, and the Center has been heavily involved in overseeing scientific payloads for the Space Shuttle. Today, the MSFC is playing a leading role in developing the Ares rockets that will take us back to the Moon.

One of the most historically significant facilities at the MSFC is the Neutral Buoyancy Space Simulator, a giant pool of water 75 feet wide and 40 feet deep. The upward buoyant force provided by the water nearly cancels out the downward force of gravity and thus provides astronauts with an environment where they can experience the apparent weightlessness of space. The astronauts' training here was essential to the success of the Gemini, Apollo, Skylab, and Space Shuttle missions. This was the only facility of its kind until the mid-1970s when a similar facility was built at the Johnson Space Flight Center.

Visiting Information

With more than 1,500 space- and rocket-related artifacts, the U.S. Space and Rocket Center is a showcase for the hardware of the U.S. space program. Inside the Space Museum, you can see simulators and trainers used for the Mercury, Gemini, and Skylab missions and the actual Apollo 16 Command Module. Several exhibits are of special note: "Rocket City Legacy" chronicles Huntsville's role in the space program; "Team Redstone" demonstrates the army's response to today's security threats; and "Earth, Moon, Mars—Living in a Challenging Environment" discusses how humans will survive on those inhospitable worlds. The museum houses several small theaters along with the Spacedome IMAX Theater where you can watch space-themed movies on the giant screen.

Outside the museum, the main attraction is the Rocket Park where you can stroll among the Hercules, Lance, Pershing, and Patriot missiles and the towering Redstone, Atlas, Jupiter, and Saturn V rockets. The first Saturn V ever built for NASA is displayed horizontally and dissected into its various stages. This is a real, fully operational Saturn V that was used for testing purposes at the MSCF. Of the three remaining Saturn Vs, this is the best preserved and is listed as a National Historic Landmark. Another outside attraction is the G-Force Accelerator, a spinning ride that lets you experience three times the force of gravity. If that's not enough, hop aboard the Space Shot where you'll be blasted 140 feet into the air with an acceleration of four "g"s.

In the Shuttle Park, you'll see a full-sized Space Shuttle complete with orbiter, boosters, and fuel tank. The Pathfinder orbiter on display here never flew in space, but it was used to test procedures and equipment. The external fuel tank was the first ever built for NASA and was used to test the shuttle's main engines. Also of significance in the Shuttle Park is the Centaur G-Prime upper stage, which has helped launch more missions to the Sun, Moon, and Mars than any other rocket.

Your admission includes a bus tour of the MSFC. Tour stops include laboratories, rocket test facilities, assembly areas for the International Space Station, and the Neutral Buoyancy Simulator. The center is also home to the

> Website: www.spacecamp.com/museum
> Telephone: 256–837–3400

U.S. Space Camp, which offers a wide variety of programs for children and teens ages 7 to 18. Some programs allow parents to participate alongside their children. See the website for details.

The U.S. Rocket and Space Center is open from 9:00 A.M. to 5:00 P.M. daily except major holidays. Admission is $18.95 for adults and $12.95 for children ages 6 to 12. The center is located at exit 15 of I-565 east of downtown Huntsville.

Space Center Houston at the Johnson Space Center, Houston, Texas

"Houston, we've had a problem." When *Apollo 13* commander Jim Lovell spoke those words, he was talking to the mission control room at the Lyndon B. Johnson Space Center (JSC) in Houston, Texas. Opened in 1963, this facility, formerly known as the Manned Spacecraft Center, has been responsible for coordinating every manned NASA mission beginning with Gemini 4 in 1964. Whenever you hear the astronauts talking to their earth-bound colleagues, they are talking to engineers and technicians here at the JSC.

The JSC encompasses more than 140 buildings on a 1,600-acre site and employs more than 19,000 NASA employees and contract workers. It serves as home base for the nation's astronaut corps, currently numbering around a hundred strong. Here, astronauts are selected and undergo their rigorous training regimen. The JSC is heavily involved in the design, development, and testing of manned spacecraft. One of the most significant testing facilities is the Space Environment Simulation Laboratory (SESL) where spacecraft hardware is subjected to the extreme thermal and vacuum conditions it will encounter in space. The testing performed at the SESL helps insure the safety of the astronauts and the success of the manned space program. Finally, the JSC houses the Lunar Receiving Laboratory, in which astronauts returning from the Moon were quarantined and where the Moon rocks and soil samples are studied and stored.

Visiting Information

Visitors to the JSC are welcomed at the Space Center Houston visitor center, designed by Walt Disney Imagineering. There's a lot to see and do here. The Astronaut Gallery holds what NASA claims is the world's best collection of spacesuits along with photos of every U.S. astronaut who has ever flown in space. To get a sense of what it's like to live in space, visit the Living in Space module where you'll see a presentation on how astronauts accomplish simple everyday tasks like eating and showering made complicated by a microgravity environment. In the Starship Gallery, you can see the film *On*

NASA

This is a view of the Mission Operations Control Room in the Mission Control Center at the Johnson Space Center during the flight of *Apollo 13*. The photograph was taken on the evening of April 13, 1970, during the fourth television transmission. The explosion occurred shortly after the end of the transmission.

Human Destiny and get a close-up look at an impressive array of space arti-facts including the actual Mercury Atlas capsule that carried Gordon Cooper into orbit, the Apollo 17 command module, a Lunar Roving Vehicle Trainer, and training apparatus for the Skylab and Apollo-Soyuz missions. In the Blast-Off Theater, you get a multisensory experience of what a launch is like, complete with clouds of exhaust gases. The Kids Space Place gives younger children a place to play.

To actually see the Johnson Space Center, be sure to take the NASA Tram Tour. On this guided tour, you'll visit the historic Mission Control Center and the Space Vehicle Mockup Facility. There's an optional stop at the Rocket Park where you can get a close look at some rockets used in the early days of the space program. Occasionally, the tram may stop at an astronaut-training facility or the new Mission Control Center.

Space Center Houston is normally open from 10:00 A.M. to 5:00 P.M. on weekdays and 10:00 A.M. to 6:00 P.M. on weekends. Extended hours are offered in the summer and on some holidays. Check the online Events Cal-endar for up-to-date information. Admission to Space Center Houston is

$18.95 for adults, $14.95 for children ages 4 to 11, and $17.95 for seniors. There is a $5 parking fee.

For the hard-core space enthusiast with deep pockets, there's the $70 Level 9 Tour. This extensive 4.5- to 5-hour tour includes several sites usually not included on the tram tour, such as the Space Environment Simulation Lab, the Teague Auditorium where more space artifacts are on display, and the Sonny Carter Training Facility-Neutral Buoyancy Lab. A lunch is included at the Astronaut Cafeteria where the astronauts eat every day. The Level 9 Tour is limited to twelve persons per day. Advance reservations are required and may be made online or by calling Group Sales at 281–244–2115. You must be 14 years of age or older to take this tour, which begins at 11:45 A.M. The $70-price

> Website: www.spacecenter.org
> Telephone: 281–244–2100

includes a two-day admission to Space Center Houston. The JSC is located at 1601 NASA Parkway about 25 miles south of downtown Houston.

New Mexico Museum of Space History, Alamogordo, New Mexico

The New Mexico Museum of Space History sits on the slopes of the Sacramento Mountains overlooking the city of Alamogordo and the Tularosa basin. Architecturally, the building consists of a five-story gold-tinted glass cube framed in concrete with a triangular concrete feature at its base. The entire building looks like a rocket ready for launch. After buying your ticket, you'll be invited to take the elevator to the top floor and work your way down through the displays. The "Icons of Exploration" exhibit takes you back to the early days of space exploration and includes displays on the Mercury, Gemini, Apollo, and Soviet programs. Replicas of the *Sputnik* and *Explorer* satellites are displayed along with a Gargoyle guided missile. Don't miss the Moon rock encased in a glass pyramid. Space suits, astronaut food, and personal hygiene equipment are among the items on display in the "Living and Working in Space" gallery. Here, you can also walk through a crew module and read about life aboard the Skylab and Salyut space stations. There is also a mock-up of the International Space Station currently under construction. The "Rockets" exhibit describes the history and purposes of rocketry. A collection of missiles is on display and you can listen to the sounds of six different rocket engines on the rocket sound board.

If you're wondering why there's a space museum in the middle of New Mexico, the "Space Science in New Mexico" exhibit will answer that question. Robert Goddard tested many of his rockets near Roswell and the captured V-2 rockets were tested at the nearby White Sands Missile Range. Other exhibits focus on meteorites, photographs from the Hubble Space Telescope, and the X-Prize/X-Cup, a competition hosted by the state of New Mexico to challenge private groups and companies to build the first commercial spacecraft. Toward the end of your descent through the museum, you'll enter the "International Space Hall of Fame." Established in 1976, this hall celebrates men and women from around the world and throughout history who have advanced our understanding of the universe and our ability to explore space. The names run the historical gamut from Galileo to Gagarin, and each inductee is recognized with a poster describing their accomplishments and contributions.

After touring the museum, go outside and wander through the John P. Stapp Air and Space Park. The park is named in honor of Dr. John Stapp, an Alamogordo resident who did pioneering research on the effects of mechanical forces on living tissue. On display in the park is the Sonic Wind No. 1, a rocket-powered sled that Dr. Stapp used to test a pilot's ability to withstand g-forces. In 1954, Dr. Stapp strapped himself into the sled and accelerated from rest to 632 mph in just five seconds, setting a new land speed record. Other rocket-related artifacts on display include the "Little Joe II," which tested the Apollo launch escape system, an F-1 rocket engine, and a variety of missiles.

Be sure to look for the sheet metal building that holds the "Daisy Track" exhibit. Named after the Daisy Air Rifle, the Daisy Decelerator was used from 1955 through 1985 to test the effects of rapid acceleration on the human body and equipment. These tests helped insure the success of NASA's Mercury program and the Apollo Moon landings. The exhibit includes the Daisy Track and Air Gun, a Waterbrake, and four original sleds. Also on the grounds of the museum is the Astronaut Memorial Garden commemorating the astronauts who died on board the Space Shuttles *Challenger* and *Columbia*. The Clyde W. Tombaugh IMAX Dome Theater and Planetarium is in a separate building nearby.

Visiting Information

The New Mexico Museum of Space History is located in Alamogordo, New Mexico. The museum is open from 9:00 A.M. to 5:00 P.M. every day except major holidays. Admission is $3 for adults, $2.75 for seniors, and $2.50 for children ages 4 to 12. Tickets for movies at the IMAX are $6 for adults, $5.50 for seniors, and $4.50 for children. There is

Website: www.spacefame.org
Telephone: 1–877–333–6589

a gift shop in the main museum building with vending machines nearby. No credit cards are accepted for payment of admission.

Virginia Air and Space Center, Hampton, Virginia

The Virginia Air and Space Center is the visitor center for NASA's Langley Research Center and Langley Air Force Base. The Langley Research Center was established in 1917 as the nation's first civilian aeronautics laboratory. More than half of the research done at Langley still focuses on aeronautics while the rest is on a variety of space-related topics, including the aerodynamics of flying through alien atmospheres. There are no public tours of the facilities at Langley, but the center makes up for this lack.

The star of the show here is the actual Apollo 12 Command Module that was launched to the Moon in November 1969. The crew consisted of astronauts Charles Conrad, Richard Gordon, and Alan Bean. The Lunar Module set down safely in an area of the Moon known as the Ocean of Storms, and Conrad and Bean became the third and fourth persons to walk on the Moon. They spent more than 31 hours on the surface and performed two separate Moon walks. They collected about 75 pounds of lunar rock and also retrieved parts of a Lunar Surveyor 3 spacecraft that had landed nearby two-and-a-half years earlier. Other spacecraft of note on display here are: a hatch from the actual Gemini 10 spacecraft, the Mercury 14 spacecraft, a full-scale model of a Viking Lander, a Lunar Orbiter, and a Lunar Excursion Module Simulator that was used at Langley.

In addition to a hangar full of air and spacecraft, the center includes an Adventures in Flight gallery, a Space Gallery, and an exhibit on ham radio. The Space Gallery boasts the largest scale models of the planets on display inside a museum in the United States. Jupiter, the largest of these heavy duty Styrofoam models, is ten feet in diameter and weighs more than 750 pounds. On this scale, the Earth is the size of a soccer ball. The "Interplanetary Travel Agency" can help you plan a trip to another world by giving you weather forecasts, travel tips, and local gravity information. In a mere nine minutes the "Mars Transporter" can take you on a trip to Mars, where you will exit onto a simulated Martian surface. The "Magic Planet Global Projector" displays images of the Earth, Moon, Sun, and Mars onto a five-foot diameter hemisphere. A Lunar Lander Simulator lets you try your hand at landing on the Moon, and you can program a Mars Rover for a mission to the red planet. A Moon rock returned by the Apollo 17 astronauts is on display along with a meteorite from Mars. Other exhibits tell about the Mercury and Gemini missions, explain rockets and satellites, and describe what the weather is like on other planets. A space-themed play area is available for little astronauts.

Visiting Information

The Virginia Air and Space Center is located in downtown Hampton, Virginia. A free parking garage is located about a block from the center. During the summer, the center's hours are Monday through Wednesday, from 10:00 A.M. to 5:00 P.M., and Thursday through Sunday, from 10:00 A.M. to 7:00 P.M. Hours during the remainder of the year are slightly more limited. Admission to the exhibits only is $9 for adults, $7 for children

Website: www.vasc.org
Telephone: 757–727–0900

ages 3 to 18, and $8 for seniors 65 and older. There is an additional charge for 3D IMAX theater tickets. See the website for a complete list of admission packages. The center has a gift shop and a "Cosmic Café."

7
Astronomers, Astronauts, and Einstein

I have loved the stars too fondly to be fearful of the night.

From *The Old Astronomer to His Pupil* by Sarah Williams

In this chapter, you meet eight American astronomers and astronauts who explained and explored space. I chose these particular people for two reasons: each made a major contribution to the field, and either a house or an entire museum in the United States offers you a place to make a special connection with the life of that person. David Rittenhouse and Maria Mitchell were astronomers in eighteenth- and nineteenth-century America who introduced the practice of science to a young nation and who were at least partly responsible for establishing a tradition of scientific excellence here. In the early twentieth century, Albert Einstein redefined our conceptions of space and time and Edwin Hubble's observations proved that space is expanding.

In the mid-twentieth century, humans began to travel into space. In the United States, we called these explorers *astronauts* from Greek words meaning "star sailor." Our Russian rivals preferred the term *cosmonaut* from Greek words meaning "universe sailor." When NASA chose the original *Mercury 7* astronauts, they had to meet certain minimum qualifications: each was younger than 40 years old and less than 5'11" tall; each held a bachelor's degree in engineering; each was a qualified jet pilot; each graduated from test pilot school; and each had logged at least 1,500 hours of flying time. In the Mercury program, 500 candidates qualified and 110 survived the initial screening. Today, the minimum requirements for a mission specialist include a bachelor's degree in engineering, physical or biological science, or mathematics. An advanced degree is preferred. You must be able to pass a NASA space physical including specific vision and blood pressure requirements, and you must be between 58.5 and 76 inches tall. Go to the NASA website for details and an application, but be prepared to face some stiff competition. Every couple of years, thousands of applicants vie for just twenty slots. The astronauts included below are Virgil "Gus" Grissom, the second American in space and among the first to give his life in the line of duty; John Glenn, the first American to orbit the Earth; Donald "Deke" Slayton, who flew on the historic Apollo-Soyuz mission; and Neil Armstrong, the first human being to set foot on the Moon.

David Rittenhouse Birthplace, Philadelphia, Pennsylvania

David Rittenhouse was a Revolutionary War patriot and the foremost American astronomer of the eighteenth century. He was born in 1732 in Germantown, Pennsylvania, a town that is now within the city limits of Philadelphia. His grandfather, William Rittenhouse, had built the first papermill in America in 1690. From an early age, Rittenhouse showed a talent for mathematics and building mechanical models, which he constructed with the aid of a box of tools inherited from an uncle. At the age of eight, he built a working model of a water mill. He had little formal schooling but instead taught himself arithmetic and geometry from books. He learned physics by reading Isaac Newton's *Principia*. When he was nineteen, he opened a shop along the road that ran through his father's farm. There he sold clocks, scientific and surveying instruments, and mechanical models of the solar system called orreries. Two of his surviving orreries are on display in the library of the University of Pennsylvania and in Peyton Hall at Princeton University. Rittenhouse, who began making astronomical instruments in 1756, is believed to have constructed the first telescope ever made in the United States.

Rittenhouse built an observatory near his home to observe a transit of Venus in 1769. Archeologists have attempted to locate the remains of this observatory but the results have been inconclusive. During the transit, Rittenhouse discovered that Venus had an atmosphere. A similar observation had been made by a Russian astronomer in 1761, but neither report was publicized for more than a hundred years.

During the Revolutionary War, Rittenhouse was called upon to use his mechanical expertise to help with the manufacture of cannons and ammunition. He served in the Pennsylvania General Assembly, on the Board of War, and as state treasurer. After the war, his attention returned to science. In the 1780s, he published papers on a variety of scientific topics including magnetism. In 1785 Rittenhouse made what was probably the world's first diffraction grating, a device that, like a prism, spreads light out into its constituent wavelengths. Rittenhouse's crude grating consisted of two very finely threaded brass screws, which formed the top and bottom of the frame. He then stretched fifty hairs across the threads forming a grating with a spacing of about one hundred lines per inch. Rittenhouse unfortunately lacked the time to pursue the idea, and, although he did publish a paper on the subject, it attracted little attention. Today, diffraction gratings are an incredibly useful tool for scientific research.

During the last years of his life, Rittenhouse published several papers in mathematics. George Washington appointed him as the first director of the U.S. Mint in 1792, and in 1795 he was elected a fellow of the Royal Society of London, which at the time was the world's leading scientific organization. He died in Philadelphia in 1796.

Visiting Information

The birthplace of David Rittenhouse, the "Rittenhouse Homestead" is part of Historic RittenhouseTown in Philadelphia. It is approximately a 20-minute drive from Philadelphia's City Hall; see the website for directions. This collection of buildings is considered to be one of the most important and best-preserved eighteenth-century industrial sites in North America and is a National Historic Landmark. The Visitor Center is open for tours on weekends from June through September. Hours are from noon until 4:00 P.M. with the last tour beginning at 3:00 P.M. Tours at other times may be arranged in advance by calling the number below. The admission fee is $5 for adults and $3 for children

Website: www.rittenhousetown.org
Telephone: 215–438–5711

and seniors. RittenhouseTown is the birthplace of paper in the United States, and the tours begin with a video appropriately entitled "The Fiber of

History." Tours include visits to the second paper mill site, the Rittenhouse Bakehouse, and the Rittenhouse Homestead (built in 1707 and thus one of the oldest homes in Philadelphia). An exhibit on the life and achievements of David Rittenhouse is planned for the future.

Maria Mitchell House, Nantucket Island, Massachusetts

Born in this house on Nantucket Island in 1818, Maria Mitchell, the first female professional astronomer in the United States, made major contributions to the education of women and women's rights. Mitchell was fortunate that her father, William Mitchell, did everything he could to encourage his daughter's early interest in science and mathematics. At the age of seventeen, she opened her own school for girls, trained them in science and mathematics, and began a lifelong passion for educating others. In 1838, Mitchell was appointed librarian at the Nantucket Athenaeum and used the book collection to further her education. The library was only open in the afternoons and on Saturday night, a schedule that provided her with ample time to study and observe the sky. Her father had built an observatory on the roof of the Pacific Bank building, where Maria often joined him in the evenings. Maria Mitchell's life changed forever when, on a crisp autumn night in 1847, she focused her father's little two-inch telescope on a spot of light that she at first assumed was a star. But the spot was too blurry to be a star, and she suddenly realized that she had discovered a comet. A few years before, King Frederick of Denmark had established gold medal prizes to be awarded to each discoverer of a telescopic comet. Mitchell duly won one of these medals, which brought her worldwide acclaim. As a result, she was the first woman to be elected to the American Academy of Arts and Sciences, the American Philosophical Society, and the American Association for the Advancement of Science.

In 1860, after the death of her mother, Maria and her father moved to Lynn, New York, where Matthew Vassar appointed her as the first faculty member at his new college for women. She served as a professor of astronomy at Vassar College from 1865 through 1888. Throughout her teaching career, Mitchell encouraged her students the same way her father had encouraged her. Her teaching methods were unconventional, eschewing the usual lecture method in favor of observation, small classes, and individual attention. As a scientist, she was much more of an observational rather than a theoretical astronomer. She was particularly interested in the Sun and traveled long distances, sometimes with a few of her students, to personally observe several total eclipses. She tracked sunspots and speculated that they were gaseous storms on the Sun's surface. She viewed Jupiter's clouds not as a mere atmospheric phenomenon, but as part of the main body of the planet itself, presaging modern thought, and she claimed that the rings of Saturn were of a different composition than the planet itself.

Throughout her life, Mitchell supported the cause of women's rights. She was friends with many leaders of the suffrage movement, cofounded the American Association for the Advancement of Women, and led a session of the women's congress. Upon learning that some of the younger, less experienced, male professors at Vassar were earning a higher salary than she was, she demanded a raise and got it. Maria Mitchell died in 1889 at the age of seventy-one. She was posthumously inducted into the National Women's Hall of Fame and, most appropriately, a crater on the Moon has been named in her honor.

The Mitchell house, built in 1790, has changed little since then. The paint, which uses a technique called "graining" so that it resembles knotty pine, is original, as are the decorative floor paintings. The house contains family heirlooms along with some of Maria Mitchell's personal belongings, including beer mugs, opera glasses, and one of her telescopes. Next door to the house is the Maria Mitchell Observatory founded in 1908. The observatory offers extensive public education programs, and undergraduate students use it to do astronomical research. The observatory with its outdoor solar system scale model, sundial, and astronomy exhibit, offers solar observing on clear days. The observatory also houses 8,000 glass photographic plates that provide a record of the night skies over Nantucket from 1910 through 1995.

Visiting Information

The Maria Mitchell House, a National Historic Landmark, is located at One Vestal Street on Nantucket Island in Massachusetts. You have to take a ferry from the mainland to get here. The Mitchell House is

Website: www.mmo.org
Telephone: 508–228–2896

open to the public from mid-June through September 1. Hours are from 10:00 A.M. to 4:00 P.M., Tuesday through Saturday. Admission is $5 for adults and $4 for children.

Edwin Hubble Home, San Marino, California

In the 1920s, astronomer Edwin Hubble redefined the cosmos by discovering that we live in an expanding universe populated by billions of galaxies. Hubble was born in Marshfield, Missouri, on the edge of the Ozark Mountains in 1889, the son of an insurance executive and lawyer. At the age of nine, he moved with his family to the comfortable Chicago suburb of Wheaton where he went to high school. The young Hubble was blessed with all the gifts that nature could possibly bestow on a single human being. He was intellectually brilliant, a truly exceptional athlete, charming, and, by all accounts, extremely handsome. At a single track meet during his senior year, he won seven events and placed third in another. Later that year, he set a state record for the high jump.

In 1906, he was awarded a scholarship to attend the University of Chicago where he studied physics and astronomy under such scientific luminaries as physicist Robert Millikan. He continued his athletic pursuits and excelled at basketball and boxing. In fact, some fight promoters tried to talk him into a career as a professional fighter. Upon graduating, he was awarded one of the very first Rhodes scholarships to attend Oxford University. But before leaving for England, Hubble promised his dying father that he would study law rather than science. He kept his promise and studied jurisprudence along with Spanish and literature. Hubble fell in love with English culture and transformed himself into an English gentleman. When he returned to Wheaton in 1913, he was wearing a cape, smoking a pipe, and speaking with an affected English accent. Hubble taught high school Spanish and mathematics and coached basketball in New Albany, Indiana, for a year, but after the death of his father, he returned to the University of Chicago to pursue graduate work in astronomy. Hubble honed his skill with a telescope at the university's Yerkes Observatory where he studied faint blobs of light called nebulae, objects that eventually made him famous.

In 1919, after military service in World War I, Hubble joined the staff of the Mount Wilson Solar Observatory where he had at his disposal the 100-inch Hooker reflector, the largest telescope in the world. Wielding the giant telescope, Hubble made discoveries that radically changed our conception of the universe. (See chapter 2, Mount Wilson Observatory, for more details.)

Hubble's discovery that the universe is expanding grabbed the attention of none other than Albert Einstein. When Einstein developed the equations describing the General Theory of Relativity, the equations actually predicted an expanding universe, yet his astronomical colleagues assured him that the universe was static and unchanging. To accommodate this view, Einstein reluctantly added a "cosmological constant"—a sort of antigravity force that would guarantee a static universe—to his equations. Einstein was elated when he heard of Hubble's discovery and pronounced the "cosmological constant" the greatest blunder of his scientific career. In 1931, Einstein visited Hubble at Mount Wilson and thanked Hubble personally for making the discovery that allowed Einstein to erase the ugly constant. Einstein's fame rubbed off on Hubble. In 1936, he published *The Realm of the Nebulae* (Hubble insisted on using the term "nebulae" rather than "galaxy"), a popular account of his discoveries that added to Hubble's growing public reputation. Tourists and Hollywood movie stars and moguls flocked to Mount Wilson to see the observatory. Hubble and his wife Grace enjoyed the company California's elite society, including Charlie Chaplin, Helen Hayes, and William Randolph Hearst.

Truth be told, Edwin Hubble had his share of personal flaws. Many of his colleagues in the astronomical community considered him arrogant and abrasive. He was also a compulsive liar who happily fabricated stories highlighting his own bravery and athletic prowess; these stories included rescuing drowning swimmers, leading soldiers to safety on the battlefields of France (Hubble arrived in France just one month before the war's end and almost certainly saw no combat), and pummeling world champion boxers. Of course, these foibles must be balanced against magnanimous acts, such as encouraging the former mule team driver and observatory janitor Milton Humason to pursue his interest in astronomy and enlisting Humason's help in making the necessary observations. Hubble gave Humason full credit in the scientific papers they published together.

Late in his life, Hubble tried to convince the Nobel Prize committee to consider astronomy as a branch of physics so that deserving astronomers, himself included of course, could be considered for the prestigious prize. (There is no Nobel Prize for astronomy.) In this Nobel quest, Hubble even went to the trouble of hiring a publicity agent to promote himself as a worthy recipient. In 1953, the Nobel committee finally agreed to consider astronomical discoveries for the Nobel Prize in physics, but it was too late for Hubble. He died a few months before the decision, and the prize is never awarded posthumously. Insiders say he was on the verge of winning. Inexplicably, Hubble's wife refused to have a funeral and did not tell anyone what she did with the body; thus, the whereabouts of the remains of one of history's greatest astronomers is unknown.

Today, Hubble's name is more famous than ever because it adorns the Hubble Space Telescope, an instrument that, like its namesake, has created an astronomical revolution. The Hubble Space Telescope has provided astronomers and all of humanity with incredible images of the beautiful and vast cosmos that Edwin Hubble first glimpsed from a California mountaintop.

Visiting Information

The Hubble House is located at 1340 Woodstock Road in San Marino, Los Angeles County, California. The house is a private residence and can be viewed from the outside only. It is listed as a National Historic Landmark.

Gus Grissom Memorial, Mitchell, Indiana

Virgil Ivan "Gus" Grissom, an original *Mercury 7* astronaut, flew into space twice on missions in both the Mercury and Gemini programs. He was training for a third trip into space when he was killed in the tragic *Apollo 1* fire. Grissom was born in the southern Indiana town of Mitchell in 1926. His father worked for the railroad, managed to avoid being laid off during the Great Depression, and earned $24 a week, enough for the family to live comfortably in their white frame house. Grissom was only an average student, although he did excel in mathematics, and he was too short to win a spot on any of the athletic teams. Instead, he was active in the Boy Scouts. He earned spending money by delivering newspapers twice a day and, in the summers, helping farmers pick cherries and peaches. He spent quite a bit of his money taking his girlfriend, Betty Moore, to the late movie. Gus met Betty during his sophomore year, and they knew right away that they were meant for each other; they married in July of 1945.

After graduation, Grissom went for training as an aviation cadet, but the war ended before he saw any combat. After a series of unsatisfying military desk jobs, Grissom left the service in November 1945. He got a job installing doors on school buses but yearned for something better. He finally decided to attend Purdue University and study mechanical engineering. Gus and Betty rented a tiny apartment off campus. To make ends meet, Gus flipped hamburgers thirty hours a week at a local diner, and Betty worked for the phone company. Grissom completed his degree in 1950 and decided to pursue his dream of becoming a pilot. He reenlisted in the Air Force, finished his training, and earned his wings. Grissom served in Korea and completed 100 combat missions, which earned him the Air Medal with clusters and the Distinguished Flying Cross. For the next several years, Grissom took various military assignments, including a stint as a flight instructor, and continued to improve his skill as a pilot. Eventually, Grissom was accepted into the test pilot school at Edwards Air Force Base and received his test pilot credentials in 1957. Grissom's life changed when he received a mysterious "Top Secret" order instructing him to report to Washington, D.C., wearing civilian clothes. When he arrived at the indicated address, he found out that he was one of 110 military pilots whose records had earned them an invitation to compete for one of seven spots as an astronaut for Project Mercury. Grissom accepted the invitation and reported to Wright-Patterson Air Force Base and Lovelace Clinic to submit to a grueling series of physical and psychological tests. He was nearly disqualified when the doctors found out that he suffered from hay fever. Grissom informed the good doctors that he was unlikely to encounter any pollen grains in outer space, and they allowed him to continue with the tests.

On April 13, 1959, Grissom received word that he had been selected as one of the Mercury astronauts. On July 21, 1961, Gus was launched into space on the second Mercury mission. All went well with the flight until splashdown when the hatch prematurely blew off the capsule allowing it to fill up with water. Grissom was hoisted to safety by a helicopter, but the Liberty Bell capsule sunk to the bottom of the Atlantic Ocean. After an exhaustive investigation, a review board found that Grissom in no way contributed to the detonation of the hatch. Nevertheless, questions lingered, and some blamed Grissom for the loss of the capsule. Grissom is unfairly portrayed in the otherwise excellent movie *The Right Stuff,* which leaves the audience with the impression that Grissom was at fault. NASA's confidence in Grissom is evidenced by his later participation in the Gemini and Apollo programs. In fact, Grissom was so heavily involved in the design of the Gemini capsule that some colleagues referred to it as the "Gusmobile." (See chapter 6, The Kansas Cosmosphere, for more details on the flight of the Liberty Bell 7.)

America's first person in space, Alan Shepard, was scheduled to command the first manned Gemini flight. Shepard, however, experienced periodic episodes of nausea, dizziness, and vomiting; he was diagnosed with a chronic inner ear disorder and grounded. Grissom, who had been Shepard's back-up, found himself in the Gemini commander's seat. He was paired with John Young, a navy test pilot who had been selected for the second group of astronauts.

As commander, it was Grissom's prerogative to name the spacecraft, and he took his inspiration from the Broadway musical *The Unsinkable Molly Brown*. He decided to poke some fun at the Liberty Bell incident and suggested the name "Molly Brown" after the title character who helped save the lives of several shipmates on the *Titanic*. His bosses balked and insisted that a more dignified name be chosen. Undeterred, Grissom suggested an alternative: the "Titanic." The NASA brass backed off and reluctantly agreed to name the Gemini 3 the "Molly Brown." After that, NASA took the naming of the spacecraft out of the hands of the astronauts.

The *Gemini 3* orbited the Earth three times in March of 1965 and the mission went smoothly except for a corned-beef sandwich that Young had smuggled aboard the spacecraft. The astronauts shared a few bites of the sandwich then stowed it away fearing that the crumbs might damage the sensitive electronic equipment. The media latched on to the "sandwich affair," and a few congressmen complained that NASA had lost control of its astronauts. Nevertheless, Grissom remained a favorite of NASA and was chosen to command the first Apollo mission.

The first Apollo vehicle, Spacecraft 012, had plenty of problems right from the start, and Grissom knew it. On his last day at home, he cut a lemon from a tree in his back yard to hang around a full-scale duplicate of the troubled spacecraft. On the evening of January 27, 1967, during a test of the Apollo spacecraft and the Saturn 1B rocket, a frayed wire under Grissom's seat sparked and ignited a fire. The flames fed on the pure oxygen gas in the capsule's interior and burned combustible materials inside the spacecraft. These materials released poisonous gases that suffocated the three astronauts before they could escape. Ironically, the multilayered hatch, which took at least 90 seconds to open, had been designed so that it would not prematurely blow open like Grissom's Liberty Bell capsule.

There are actually two Gus Grissom Memorials, one in his hometown of Mitchell and the other in nearby Spring Mill State Park. Near downtown Mitchell, on the spot once occupied by his elementary school, stands Mitchell's Gus Grissom Memorial, a carving of the Gemini rocket in Indiana limestone. At the base of the monument eight panels tell Grissom's life story. The monument is surrounded by a wall made from bricks from the school. A nearby marker reads: "One of Ours—In grateful tribute to Virgil I.

Grissom—Master Mason—Explorer and trailblazer in the best of American traditions." The municipal building on the other side of the parking lot has a small display about Grissom. A few blocks away Grissom's childhood home is being converted into a museum.

The Gus Grissom Memorial in Spring Mill State Park was dedicated in 1971 four years after his death. The main attraction here is the actual Gemini 3 "Molly Brown" capsule. It is accompanied by artifacts from the capsule including the hatch, Gus's Gemini spacesuit, and his helmet. Grissom's helmet from the Mercury mission is also on display. A 15-minute video on his life can be viewed in the mini-theater.

Visiting Information

Mitchell is in southern Indiana, about a 90-minute drive from Indianapolis. Mitchell's Grissom Memorial is located on 6th Street. Grissom's home is at the corner of 8th Street and Gus Grissom Avenue. The Gus Grissom Memorial at Spring Mill State Park is off of Highway 60 about three miles east of Mitchell, Indiana. Hours are 9:00 A.M. to 4:30 P.M., and there is a $7 entrance fee to the park for out-of-state residents. If you are just going to the memorial, tell the guard, ⌐Telephone: 812–849–4129⌐ who might allow you in for free. (The park's other worthwhile sites include a "Pioneer Village" with a restored grist mill and boat rides into a cave.) A special Saturday celebration called "Gus Grissom Day" is scheduled in the park during the summer.

John and Annie Glenn Historic Site, New Concord, Ohio

John Glenn—pilot, astronaut, and U.S. senator—is the iconic All-American hero. He was born in Cambridge, Ohio, in 1921, but at the age of two he moved with his parents to New Concord. The home had extra rooms that were occupied by college students from nearby Muskingum College. John, encouraged by a father who enjoyed travel and tutored by a loving mother, was surrounded by college students who modeled an academic lifestyle. In this enriching environment John developed an early interest in science and a fascination with flying. Glenn later wrote that: "A boy could not have had a more idyllic early childhood than I did." He met his future wife Annie when they were both toddlers. The parents of John and Annie were good friends and often had dinner together. On these occasions, John and Annie shared a playpen. In his autobiography, Glenn poignantly writes: "She was part of my life from the time of my first memory." Their relationship evolved from childhood playmates to high school and college sweethearts. They both graduated from Muskingum College, John with a degree in engineering and Annie with a degree in music; they were married in 1943.

Annie Glenn has struggled with a severe stuttering problem all of her life. In 1973, she participated in an intensive experimental treatment program and now speaks fluently and even makes public speeches. Annie has been active in promoting programs for children, the elderly, and the handicapped. In 1983, she was the first recipient of a national award from the American Speech and Hearing Association for "providing an inspiring role model for people with communicative disorders."

In March 1942, Glenn enlisted in the Naval Aviation Cadet Program and earned his wings as a marine pilot. During his World War II service, he flew 59 combat missions in the Pacific theatre. In the Korean conflict, he flew another 63 missions, and in the last nine days of fighting, he shot down three MiGs over the Yalu River. He received numerous combat decorations, including the Distinguished Flying Cross six times. His flying buddies noticed that Glenn had an unlucky tendency to attract enemy gunfire. On two occasions, they counted more than 250 holes in his plane after he returned from a mission, a feat that earned him the unflattering nickname "Magnet Ass."

After Korea, Glenn attended Test Pilot School, and after graduation he joined the Naval Air Test Center's elite group of expert pilots. In 1957, Glenn set a transcontinental speed record flying from Los Angeles to New York in three hours and twenty-three minutes. This feat established Glenn as one of the country's top test pilots.

When the call went out for an astronaut corps in 1958, Glenn eagerly volunteered, and the newly formed NASA chose Glenn as one of the original Mercury 7 astronauts. On February 20, 1962, John Glenn flew the "Friendship 7" capsule into space and around the Earth three times on a five-hour voyage. He was the third American astronaut in space and the first to orbit the Earth. During the flight, a signal—later found to be false—indicated that the heat shield was no longer locked into position. If that were true, the only things holding the heat shield in place were the straps of the retro pack. On reentry, Glenn was instructed not to jettison the retropack, but to leave it on to insure that the shield stayed in place. After lagging behind the Soviet Union in the Space Race, Glenn's orbital flight provided a much needed boost to American confidence and morale. He was welcomed back to Earth as a national hero and honored with a ticker-tape parade.

Glenn learned that NASA had no plans for him to fly again, but he continued to serve as an advisor at NASA until 1964, when his attention turned to politics. Encouraged by Bobby Kennedy, Glenn ran for the U.S. Senate seat in Ohio, but was forced to withdraw after falling in his bathtub and receiving a severe concussion. After a stint as a business executive, he ran again for the Senate in 1970, but this time was defeated in the Democratic primary by Howard Metzenbaum. Four years later, he tried for a third time and won handily. During his Senate career, he was active in arms control and was the chief author of the 1978 Nonproliferation Act. He sat on the Foreign Relations Committee, the Armed Services Committee, and the Committee on Aging and chaired the Government Affairs Committee. He was a contender for the vice-presidential nomination three times and ran for president in the 1984 Democratic primaries.

In 1998, a year after his retirement from the Senate, NASA invited him to return to space as a crew member of the Space Shuttle *Discovery*. Although some criticized Glenn's involvement as a publicity stunt, NASA defended it by saying they were investigating space flight and the aging process. On October 29, 1998, more than 36 years after his first space flight, John Glenn, at age 77, became the oldest human to ever venture out into space.

Visiting Information

The John and Annie Glenn Historic Site is a living history museum that shows what life was like on the home front during World War II. The one-hour tours begin with a short film. Visitors are then led through the house by costumed actors portraying various characters in the Glenn household. On the tour, you will see Glenn's boyhood bedroom along with a toy room holding his actual tricycle and train set. Other displays include a 1/3-scale model of the Friendship 7 Mercury capsule, the jumpsuit he wore after splashdown, and a few of his military uniforms. After the tour, there is a small gift shop for browsing.

The site is located in New Concord, Ohio, just north of Interstate 70 at exit 169 and is open from mid

Website: www.johnglennhome.org
Telephone: 740–826–0220

April through mid October. Hours are Wednesday through Saturday from 10:00 A.M. to 4:00 P.M. and Sunday from 1:00 P.M. to 4:00 P.M. Admission is

$6 for adults, $5 for seniors, $3 for students, and children under 6 are free.

Deke Slayton Memorial Space and Bike Museum, Sparta, Wisconsin

Donald "Deke" Slayton, another of the original Mercury 7 astronauts, was born on a farm near Sparta in western Wisconsin in 1924. Only a month after his eighteenth birthday, Slayton signed up with the U.S. Army Air Corps and learned to pilot an aircraft despite having lost his left ring finger in a farm equipment accident. During World War II, he flew 56 combat missions over Europe and another 7 over Japan. After the war, Slayton earned a degree in aeronautical engineering from the University of Minnesota. Eventually, he became a test pilot at Edwards Air Force Base, where he tested a number of air force fighters including the F-105, the principal fighter bomber used during the Vietnam War.

In 1959, Slayton was selected as one of the original seven astronauts for NASA's Mercury program. Originally scheduled to fly the fourth Mercury mission, he was grounded because of an erratic heart rate. As a result, Slayton was the only one of the Mercury 7 astronauts who did not fly a Mercury mission. In 1963, NASA picked Slayton to be "Director of Flight Crew Operations," better known, unofficially, as "Chief Astronaut." In this capacity, he was responsible for making the final decisions on crew assignments for the Gemini and Apollo programs. Slayton was finally granted medical clearance and got his chance to fly into space on the Apollo-Soyuz mission in 1975. This mission, the first meeting of American astronauts and Soviet cosmonauts, was the last flight of the Apollo spacecraft. Upon his return from space, Slayton was in charge of various testing programs for the Space Shuttle. He retired from NASA in 1982 and founded Space Services, Inc., a company that developed rockets for launching small commercial payloads into space. Slayton died of complications from a brain tumor in 1993.

The museum portrays Slayton as a brave, determined, All-American follow-your-dreams kind of guy. His personal motto was "Decide what you want to do, then never give up until you've done it." The exhibits highlight his achievements during his NASA years and describe the political, social, and scientific aspects of the Apollo-Soyuz mission. Artifacts include Slayton's unused Mercury spacesuit, an Apollo "space couch," original flight checklists, and various awards, proclamations, and letters signed by presidents from Nixon through Reagan. Kids will enjoy the museum's playroom with all kinds of puzzles, games, and videos about space. Because Sparta claims to be the bicycling capital of America, the other half of the museum is dedicated to the evolution of the bicycle.

Visiting Information

The museum is located near downtown Sparta, on the corner of North Court and West Main streets. The Space and Bike Museum occupies the second floor of the building. During the summer, museum hours are Monday through Saturday from 10:00 A.M. to 4:30 P.M. and from 1:00 P.M. to 4:00 P.M. on Sunday. During the winter,

> Website: www.dekeslayton.com
> Telephone: 888–200–5302

hours are Monday through Friday from 10:00 A.M. to 4:00 P.M. Admission is $3 for adults and $1 for children ages 6 to 13. The world's largest high wheeler bicycle can be seen at the corner of Wisconsin and Water streets.

Armstrong Air and Space Museum, Wapakoneta, Ohio

The Armstrong Air and Space Museum chronicles the history of aviation and spaceflight with an emphasis on the contributions made by Ohioans, especially Neil Armstrong, the first human being to set foot on another world. Armstrong was born on his grandparent's farm near Wapakoneta, Ohio, in 1930. His father had a job in state government that required frequent moves. The Armstrong family lived in sixteen different Ohio towns during the next fourteen years and finally returned to Wapakoneta in 1944. At age six, young Neil flew aboard a Ford Tri-Motor "Tin Goose" and fell in love with flying. He worked at a variety of jobs around town and at the airport to earn money for flying lessons. On his sixteenth birthday, Armstrong received his pilot's license. He became interested in the science of flight and built a small wind tunnel in his basement so he could test the aerodynamic properties of the model planes he built. Armstrong attended Purdue University on a navy scholarship that required him to go to school two years, then complete three years of active duty with the navy, and then return to finish his degree. During his active duty, he earned his wings at the Pensacola Naval Air Station and flew 78 combat missions during the Korean War. Upon graduating, Armstrong joined the National Advisory Committee for Aeronautics (NACA) as a research pilot. After a brief stint at the Lewis

Flight Propulsion Laboratory in Cleveland, Armstrong transferred to the NACA's High Speed Flight Station at Edwards AFB, California. There he flew more than 900 flights in more than 50 different pioneering high-speed aircraft, including 7 flights in the 4,000 mph X-15 rocket plane.

In 1962, NASA announced it was accepting applications for the second group of astronauts dubbed by the press as the "New Nine" following the original "Mercury Seven" astronauts. Armstrong applied and made the cut. His first space flight was in 1966 aboard Gemini 8. During the flight, Armstrong and David Scott performed the first rendezvous and docking of two vehicles in space by linking together the Gemini 8 spacecraft and an unmanned Agena rocket.

In the Apollo program, the normal crew rotation scheme assigned Armstrong as the commander of Apollo 11, the mission that would land on the Moon. Armstrong would be joined by lunar module pilot Buzz Aldrin and command module pilot Michael Collins. As to who would be the first to walk on the Moon, the official explanation was that the lunar module hatch opened inward to the right making it difficult for Aldrin, who would sit on the right, to exit first.

As Armstrong and Aldrin drifted down to the Moon, their main objective was to land safely rather than at some particular spot. Although most accounts report only a few seconds of fuel left upon landing, a post-mission analysis indicated that about 50 seconds of hovering time was still available. Moreover, Armstrong knew that the Lunar Module could survive a fall from as high as fifty feet. On July 20, 1969, the Lunar Module landed on the Moon at a place called the Sea of Tranquility. The original plan called for Armstrong and Aldrin to rest before exploring, but Armstrong requested that the "extra-vehicular activity" be moved up. While waiting to exit the spacecraft, Armstrong thought about what he would say when he stepped off the ladder and on to the Moon. The Lunar Module was depressurized, the hatch was opened, and Armstrong climbed down the ladder. At 10:56 Eastern Daylight Time, Neil Armstrong planted his left foot on the surface of the Moon and spoke these familiar words: "That's one small step for man, one giant leap for mankind."

These words have stirred controversy because, if you think about it, the statement doesn't make any sense. The words "man" and "mankind" are synonymous, rendering the sentence self-contradictory. Armstrong meant to say "That's one small step for a man, one giant leap for mankind." Funny how such a little word can make such a big difference. (If one listens to recordings, Armstrong pauses unnaturally between the words "one" and "giant." This suggests to me that he may have realized his mistake, but it was too late to go back.) Armstrong himself admits that he sometimes drops syllables when speaking and has been quoted as saying that he "would hope that history would grant me leeway for dropping the syllable and understand that it was certainly intended, even if it wasn't said—although it might actually have been." In 2006, an Australian computer programmer claimed that he had performed an acoustic analysis of the Apollo 11 recording and recovered the missing "a." The "a" had been rendered inaudible by the communications equipment that was used at the time. The jury is still out on this claim. Armstrong prefers that the quotation include the word "a" in brackets.

Armstrong and Aldrin spent about two-and-a-half hours exploring the surface of the Moon. They began with some ceremonial tasks, which included planting an American flag, unveiling a commemorative plaque, and taking a congratulatory phone call from President Richard Nixon. Their scientific tasks involved collecting rock samples, setting up a scientific experiment package, and seeing how easily a human could move and maneuver on the Moon. Before climbing back into the LM, the astronauts left a small memorial package in honor of dead Soviet and American astronauts. Back inside the LM, they discovered that they had accidentally broken off the switch that fired the ascent engines. Disaster was averted by using a part of a pen to close the circuit. The LM blasted off the Moon, rendezvoused with the Command Module, and safely returned to Earth. Later, Armstrong admitted that he thought the Apollo 11 mission only had about a 50 percent chance of landing on the Moon, saying that "I was elated, ecstatic, and extremely surprised that we were successful."

Armstrong left NASA in 1971 to accept a faculty position in the engineering department at the University of Cincinnati. He left that post in 1979 to serve as a spokesman for various businesses including Chrysler Corporation. He served on two spaceflight accident investigation panels, the first for *Apollo 13* and later for the Space Shuttle *Challenger*. Today, Armstrong lives in Ohio and is chairman of CTA, Inc., a computing systems corporation.

Armstrong has had a tendency to avoid the public spotlight. In contrast, his more extroverted *Apollo 11* partner, Buzz Aldrin, is regularly seen on television and in the media. Armstrong has jealously guarded his name and image. He quit signing autographs in 1994 because he was bothered by the fact that items bearing his signature were selling for thousands of dollars. In 1994, he sued Hallmark Cards after they used his name and a recording of his "one small step" quote for a Christmas ornament. Armstrong donated the settlement money to Purdue. His most bizarre legal battle occurred in 2005 when he discovered that his barber of 20 years had sold some clippings of his hair for $3,000. The barber agreed to donate his profits to a charity designated by Armstrong.

The museum named in Armstrong's honor (I hope they got his permission) is architecturally interesting; most of the building is buried under mounds of earth with the white dome of the Astro Theater towering above the main entrance. The concrete structures on either side of the dome resemble wings. The highlight of the museum's collection is the Gemini VIII spacecraft flown by Armstrong and Scott in 1966. Armstrong's Gemini and Apollo spacesuits are on display along with a Moon rock collected on the Apollo 11 mission. Other Apollo artifacts include a lunar drill, space food, in-flight suits, and a seismic experiment package. Objects from Armstrong's career as a pilot accompany the 1946 Aeronca 7AC Champion that Armstrong learned to fly in. Other exhibits of note include an airframe from the early airship Toledo, a reconstructed Wright Model G Aero-boat; a collection of model planes illustrating the rapid evolution of airplane design; and objects belonging to cosmonaut Yuri Gagarin, the first human in space. You can test your mettle for landing the Space Shuttle or the Lunar Module by climbing aboard the museum's landing simulators. The museum also holds a portrait gallery for all twenty-four Ohio astronauts.

One of the most fascinating displays concerns the so-called "Wow!" signal, famous among radio astronomers and SETI (Search for Extraterrestrial Intelligence) enthusiasts. The signal was detected by the "Big Ear" radio telescope, operated by the Ohio State University until it was torn down by land developers in 1998. On the computer printout containing the signal, its discoverer, Dr. Jerry Ehman, circled the signal with a red pen and wrote "Wow!" in the margin. It is possible that the signal was a beacon from an extraterrestrial civilization, but, alas, repeated attempts to find the signal again have been in vain. In science, if an observation can't be repeated, then it's not reliable.

Visiting Information

The Neil Armstrong Air and Space Museum is in Neil Armstrong's hometown of Wapakoneta, in western Ohio north of Dayton. The museum is located just west of I-75 at Exit 111 and is open Tuesday through

Website: www.ohiohistory.org/places/armstron/
Telephone: 1–800–860–0142

Sunday except for Thanksgiving Day. Hours are from 9:30 A.M. to 5:00 P.M. Tuesday through Saturday and from noon until 5:00 P.M. on Sunday. Admission is $7 for adults and $3 for children ages 6 to 12.

Armstrong Statue, Purdue University, West Lafayette, Indiana

Purdue has produced a plethora of astronauts (the current count is 22) including Gus Grissom, Roger Chaffee, and Gene Cernan. But the most famous is, of course, Neil Armstrong who entered Purdue University in 1947 to study aeronautical engineering. He had been accepted at MIT, but an engineer friend of his advised him that he didn't have to go all the way to Massachusetts to get a good education. Purdue dedicated the Neil Armstrong Hall of Engineering in 2007. Inside the atrium of the building is a replica of the Apollo 1 capsule in which Purdue alumni Grissom and Chaffee died in 1967. In front of the building is an eight-foot-tall bronze sculpture of Armstrong. The statue portrays Armstrong not as an astronaut, but, appropriately, as a student. He is shown wearing a button-down Oxford shirt, cuffed khaki pants, penny loafers, and a windbreaker. His right hand rests on a stack of books accompanied by a slide rule. Designed with input from Armstrong, this is the way he sees himself. As he told the National Press Club in February 2000, "I am, and ever will be, a white-socks, pocket-protector, nerdy engineer, born under the second law of thermodynamics, steeped in steam tables, in love with free-body diagrams, transformed by Laplace, and propelled by compressible flow." Armstrong is sitting on a stone plinth and is looking off to his left where an elliptical stone arc, suggestive of the trajectory of a spacecraft, is embedded in the ground. An inscription in the arc bears Armstrong's words: "One small step for a man, one giant leap for mankind." The arc leads toward a trail of twenty white boot prints set into the grass, a symbol of his historic first steps on the Moon. The impressions were molded from an actual Moon boot, and the prints are spaced far apart to mimic the stride of a bounding Apollo astronaut. The 1,800-pound statue is the work of internationally known artist Chas Fagan, who has created art for the National Cathedral in Washington, D.C., and is currently working on a statue of Ronald Reagan that will reside in the U.S. Capitol Building.

Visiting Information

The Neil Armstrong Hall of Engineering is located at Stadium and Northwestern

Website: www.engineering@purdue.edu/ armstronghall

Telephone: 765–494–5345

avenues on the campus of Purdue University in West Lafayette, Indiana, about a one-hour drive north of Indianapolis.

Einstein's House, 112 Mercer Street, Princeton, New Jersey

Albert Einstein lived in this house from 1935 until his death in 1955. When the Einsteins first arrived in Princeton in 1933, they rented a house, but, in the summer of 1935, the house down the block at 112 Mercer Street went up for sale. The home is a modest two-story white clapboard structure with a big front porch and a little front yard. The house is a reflection of the man—simple and unpretentious with a slightly disheveled look to it. Einstein lived here with his wife, Elsa, his stepdaughter, Margot, and his secretary, Helen Dukas. Through the years, a menagerie of pets kept them company, including a parrot named Bibo, a cat named Tiger, and a white terrier named Chico, who occasionally provoked unfriendly encounters with the mailman. Inside, the living room was dominated by massive German furniture. Helen Dukas converted a small library on the ground floor to an office where she could sort through letters and answer telephone calls. Einstein had a study on the second floor with a picture window that overlooked the backyard garden. Built-in bookcases climbed from floor to ceiling. The walls were graced with pictures of Einstein's scientific heroes: Isaac Newton, Michael Faraday, and James Clerk Maxwell along with a political hero, Mahatma Gandhi. The only award on display was a framed certificate of membership in the Bern Scientific Society. A large wooden table, a dumping ground for Einstein's papers and pencils, occupied the center of the room. Einstein usually worked there in the afternoons while sitting in an easy chair and scribbling an endless stream of equations onto a pad of paper balanced on his lap.

How did Einstein wind up in Princeton? Einstein was a professor in Berlin when Hitler came to power in 1933; however, Einstein was Jewish, and the Nazis wasted no time in dismissing relativity as an erroneous Jewish theory and throwing Einstein's books on their bonfires. Einstein decided to leave Germany forever and immigrate to the United States. He was drawn to Princeton to work at the newly established Institute for Advanced Study, a "think-tank" where he would receive a generous salary but remain unencumbered by teaching or research duties. He was completely free to follow his own intellectual desires—the perfect place for the free-spirited Einstein. During his time at the Institute, Einstein worked on a unified field theory, an attempt to combine the forces of gravity and electromagnetism under one mathematical framework. He never achieved this elusive goal. However, some of the world's leading physicists who work at the Institute today are continuing Einstein's quest.

Einstein died in 1955 at the age of 76. He disliked the cult of personality that had formed around him and forbade any type of funeral service. His body was cremated in Trenton on the afternoon of his death, and his ashes were scattered in the Delaware River before the rest of the world knew he was gone. He specifically requested that his house not be turned into a museum. After his death, his secretary Helen Dukas and his stepdaughter Margot continued to live in the house for more than 30 years. (His wife Elsa had died a short time after they moved into the house.) When Margot died in 1986, the house became the property of the Institute. Institute officials used the house to help entice a distinguished professor to come work here. That must have been some sales pitch: "You come work for us, and you can live in Einstein's house." It worked—the professor still lives there.

Visiting Information

The house at 112 Mercer Street is a private residence. It is not open to the public. However, the Bainbridge House at 158 Nassau Street, home of the Princeton Historical Society, has a collection of furniture that furnished the house on Mercer Street. The collection includes Einstein's favorite tub armchair, his family's grandfather clock, his music stand, and his pipe. The collection is located in the "Einstein Room" on the second floor. The Bainbridge House is the city's unofficial visitor's center and distributes free self-guided walking tour brochures and pamphlets on Einstein. Two-hour, two-mile guided walking tours are offered on Sunday afternoons at 2:00 P.M. for $7. The Bainbridge House is open on Tuesday through Sunday from noon until 4:00 P.M. Though the Institute for Advanced Study is not affiliated with Princeton University, the Institute's original offices were on the Princeton campus. From 1933 until 1939, Einstein's office was 109 Fine Hall (now Jones Hall), and he also worked in the Palmer Physics Laboratory (now the Frist Campus Center). Today the Institute for Advanced Study is located about a mile from the center of town on Einstein Drive. The buildings are

Websites: www.princetonhistory.org
www.ias.edu (Institute for Advanced Study)
Telephone: 609–921–6748 (Bainbridge House)

not open to the public, but you can at least park your car and take a look. The Institute Woods are open to the public so feel free to take a leisurely stroll.

Einstein Memorial, Washington, D.C.

Whenever I'm in Washington, I, like a pilgrim to Mecca, am inexorably drawn to this delightful memorial. Of all the monuments and memorials adorning our nation's capital, this is my personal favorite. Tucked away in a tranquil grove of elm and holly trees on the grounds of the National Academy of Sciences, the memorial is a 12-foot tall bronze statue of Albert Einstein, sitting on the second of three semicircular steps of white granite dug from the hills of Mount Airy, North Carolina. This bronze Einstein is dressed simply, as he often was in real life, in a sweatshirt and sandals. Sculpted by

The Albert Einstein Memorial in Washington, D.C.

Robert Berks, the sculpture is rendered in an artistic style, reminiscent of the impressionist paintings of Vincent Van Gogh, where a coherent whole emerges from a myriad of individual brushstrokes. (Berks also sculpted the bust of John F. Kennedy that graces the Kennedy Center for the Performing Arts.) In sharp contrast to the formality of most of Washington's monuments, this sculpture exudes a feeling of relaxed informality, which captures the essence of Einstein's personality: unassuming and unimpressed with his own importance.

In his left hand, Einstein holds a tablet with equations that summarize three of his most important scientific achievements. The top equation is the equation for general relativity, the theory that introduced a new conception of gravity; the middle equation describes the photoelectric effect, a phenomenon which Einstein explained by envisioning light as behaving like little bundles of energy called "quanta" (it was for this explanation that Einstein won the 1921 Nobel Prize in physics); the third equation is the famous $E = mc^2$ formula that sets forth the equivalence of matter and energy. Einstein seems to invite the visitor to climb into his lap and contemplate the equations. Go ahead . . . there are no security guards to chase you away. I'm sure Einstein would have liked that!

At Einstein's feet lay a 28-foot-diameter circular map of the sky made of emerald-pearl granite from Norway. The 2,700 metal studs embedded in the granite represent the positions of the planets, stars, and various other celestial objects at noon on April 22, 1979, the day the memorial was dedicated in celebration of the centennial of Einstein's birth. The size of the studs corresponds to the apparent brightness of the objects as seen from the Earth. Along the back of the steps three quotations capture Einstein's love of freedom, nature, and truth:

> As long as I have any choice in the matter, I shall live only in a country where civil liberty, tolerance, and equality of all citizens before the law prevail.

> Joy and amazement of the beauty and grandeur of this world of which man can just form a faint notion . . .

> The right to search for truth implies also a duty; one must not conceal any part of what one has recognized to be true.

Visiting Information

The Einstein Memorial is on the grounds of the National Academy of Sciences just across the street from the Vietnam War Memorial. While you are there, you might want to take a look inside the National Academy building, which houses some neat art and architecture.

Website: www.nasonline.org. Click on "About the NAS" and then "NAS Building."

8
Aliens?

A sad spectacle. If [the stars] be inhabited, what a scope for misery and folly.
If they be not inhabited, what a waste of space.   Thomas Carlyle

Are we alone? Are we, the species called *Homo sapiens,* the only intelligent life form in the universe? Nobody knows the answer to that question; we can only speculate. If a poll were taken of the scientific community, a solid majority would probably answer: "No, we're not alone." But nobody knows for sure.

There are good reasons for thinking that intelligent life might exist elsewhere in the cosmos. First, we can play the numbers game. According to the latest estimates based on photographs of deep space taken by the Hubble Space Telescope, more than a hundred billion galaxies are in the universe; each contains, on average, about a hundred billion stars. It's just hard to believe that in all this immensity of space, we're the only intelligent beings. If we are alone, then, as Thomas Carlyle first observed, what a waste of space.

A second piece of circumstantial evidence exists: complex organic molecules, molecules based on the element carbon upon which life as we know it is based, appear to be commonplace throughout the universe. Radio telescopes have found dozens of different organic molecules in the clouds of gas and dust that populate the cold, inhospitable environment of interstellar space. Organic molecules have also been found in meteorites. Although no one suggests that these organic molecules have a biological origin, their mere presence shows that the chemicals of life can be found just about anywhere.

A final fact suggesting that life might be scattered across the universe is the result of a fascinating experiment done in the early 1950s at the University of Chicago by Stanley Miller, a graduate student working with the Nobel Prize–winning chemist Harold Urey. In the experiment, the gases believed to be present in the Earth's primordial atmosphere—hydrogen, water vapor, methane, and ammonia—were mixed together in a spherical glass flask along with liquid water. Electrical sparks were applied to the mixture to simulate lightning. After allowing the experiment to run for a couple of weeks, a thick brown ooze coated the inside of the flask. A chemical analysis of the ooze revealed that it consisted of a rich variety of complex organic molecules, including amino acids. The experiment has been repeated using a variety of energy sources, including ultraviolet light and heat, but the result is always more or less the same. The chemicals and energy sources used in the Miller-Urey experiment and others like it are not unique to the early Earth. The starting ingredients are some of the most common molecules in the entire universe. These experiments hint that the basic chemical building blocks of life are easy to make.

It is one thing to argue that intelligent life exists elsewhere in the universe; it is something else entirely to claim that beings from another world have actually visited the Earth. The question "Are we alone?" must be asked separately from the question "Have they come here?" An overwhelming majority of the scientific community would answer "no" to this question. Polls, however, show that a sizable fraction of the American public believes that we are being visited by extraterrestrials. For example, a 2004 Gallup Poll found that 34 percent of Americans believe in UFOs. How do you explain this schism between scientific opinion and public opinion? Perhaps scientists tend to base their opinions on evidence, but the general public tends to base opinion on emotion. Scientists are trained to be skeptical; to withhold belief in the absence of evidence. As the late astronomer Carl Sagan said: "Extraordinary claims require extraordinary evidence." Actually, the second "extraordinary" need not be included in this statement. The standards of evidence are not any higher for "extraordinary" claims than they are for "ordinary" claims. Still, in science the point remains: claims require evidence. Unfortunately, there isn't a scintilla of credible scientific evidence supporting the claim that Earth is being visited by extraterrestrials. I say "unfortunately" because scientists would welcome such evidence. Contact with intelligent beings from another world would be the greatest scientific discovery of all time. Any scientist salivates at the chance to interact with an ET. It's not that scientists are close minded and just don't want to believe in aliens; it's that there's just no evidence. What about all the photos and videos of flying saucers? Photos and videos can be faked. In fact, my students do a project in which they must fake an encounter with a pseudo-scientific phenomenon like a UFO and produce a photograph or a videotape to corroborate their encounter. Some of their pictures look pretty convincing. It's easy to fake a photo of a UFO. Just take a disk-shaped object like a hubcap, a pie pan, or a Frisbee, toss it into the air, and snap a picture of it. What about eyewitnesses? Eyewitness testimony is unreliable in both a court of law and the scientific arena. People make mistakes, misinterpret what they see, hallucinate, and, yes, sometimes they lie to claim their fifteen minutes of fame. Anecdotal evidence is worthless in science.

The modern flying saucer craze began on June 24, 1947, when Kenneth Arnold, a part-time search-and-rescue pilot, reported seeing nine strange objects near Mount Rainier, Washington, while searching for a missing military aircraft. He later described the motion of the objects as erratic "like a saucer if you skip it across the water"; thus, the term "flying saucer" was born. Arnold's unusual encounter was widely reported by the press, and in the months that followed, thousands of similar sightings, including the now famous Roswell incident, were reported from across the country and around the world.

The term UFO stands for Unidentified Flying Object. Are there objects or lights in the sky that can't be identified? Of course! But that doesn't mean that the unknown objects must be spacecraft from an extraterrestrial civilization. Consider the necessary line of reasoning: observation—there's a light in the sky, and I don't know what it is; conclusion—it must be a spaceship full of aliens. That's quite leap of logic, don't you think?

In scientific thinking, a dictum known as Occum's Razor holds that the simplest explanation for a phenomenon is most likely to be correct. Let's apply Occum's Razor to the UFO phenomenon. Is the simplest explanation for an unknown light in the sky an alien spaceship? Surely not. Aircraft, astronomical objects such as meteors or the planet Venus, unusual weather-related phenomena, someone playing a practical joke—all of these are simpler, less exciting, more down-to-earth explanations that are also more likely to be true.

Some proponents of the idea that UFOs are alien spaceships would have us believe that there is a vast government conspiracy to cover up the evidence of alien visitation. But the government isn't very good at keeping secrets in an open society with a free press. Consider how Richard Nixon's denial of a cover-up of the Watergate break-in was undermined by the White House tapes or how Bill Clinton's denial of a sexual liaison with an intern was undermined by a stain on a blue dress. If the government were in fact trying to cover up the fact that we were being visited by aliens, don't you think somebody, somewhere, in the 60 or so years since the first flying saucers were sighted, would have gone to the press with a similar piece of incontrovertible evidence, a captain's log, say, or some alien DNA, proving an alien encounter? And yet, the only evidence is a bunch of fuzzy photographs.

Why do people persist in believing in UFOs in spite of a total lack of convincing evidence? In part, believing is simply more fun than not believing. I like to call this the "Santa Claus Syndrome." Remember when you were a little kid and believed in Santa Claus? Wasn't it a lot of fun to believe that a portly red-suited elf from the North Pole came to your house to deliver a bag full of toys? When you figured out that Santa was just pretend, wasn't that revelation tinged with just a touch of sadness and disappointment? Maybe UFOs and other pseudoscientific phenomena are adult versions of Santa Claus. Perhaps these beliefs satisfy some deep-seated psychological desire to believe in something.

Below, I describe the two of the best known UFO-related sites in the United States: Roswell, New Mexico, site of a supposed crash of an extraterrestrial craft in 1947, and Area 51, a top-secret air base north of Las Vegas where some claim an alien spaceship (the most popular theory is that the ship is the one recovered at Roswell) is stored along with several dead aliens.

Roswell, New Mexico

Roswell is the site of the most famous UFO incident in history. On June 14, 1947, rancher W. W. (Mac) Brazel found some tinfoil, wooden sticks, tape, and other debris scattered around on the J. B. Foster sheep ranch, 85 miles northwest of Roswell, New Mexico. Brazel later said he was in a hurry and just really didn't think much of it. Ten days later, the Kenneth Arnold incident started the modern flying saucer craze. Over the next several weeks, newspapers across the country reported hundreds of flying saucer sightings.

Brazel didn't learn of these reports until he drove to the nearby town of Corona on July 5, where he may have also heard a rumor about a reward for anyone who recovered one of the flying disks. On July 7, Brazel drove to Roswell and told the sheriff that he may have found a saucer. The sheriff, in turn, called Major Jesse Marcel, an intelligence officer at the Roswell Army Air Field (RAAF). Brazel escorted Marcel to the crash site, where Marcel gathered up about five pounds of scattered debris, dumped it into the trunk of his car, and drove back to the air field. After inspecting the debris, the RAAF commander, Colonel William Blanchard, issued a press release stating that a flying saucer had been recovered. The next day, the headline across the front page of the *Roswell Daily Record* read: "RAAF Captures Flying Saucer on Ranch in Roswell Region," and the sensational news spread around the world. The wreckage was shipped immediately to the 8th Air Force headquarters in Fort Worth, where Brigadier General Roger Ramey took another look at the material. After consulting with his weatherman, Ramey called in the local press and declared that the debris was from a high-altitude weather balloon, not from an alien spacecraft. On July 9, the headline for the Roswell newspaper proclaimed: "General Ramey Empties Roswell Saucer." The explanation seemed to satisfy everyone, things quickly returned to normal and, for the next thirty-two years, nothing newsworthy came out of Roswell.

That all changed in 1980 with the publication of *The Roswell Incident*, written by UFOlogists Stanton Friedman, William Moore, and Charles Berlitz (in 1975 Berlitz had written the pseudoscientific classic *The Bermuda Triangle*). The impetus for the Roswell book was an interview that Friedman had conducted with Jesse Marcel, who still held fast to the belief that the wreckage he had retrieved came from an alien spacecraft. The book claimed that the U.S. government had covered up the real story of the Roswell crash. An avalanche of books and films followed. In fact, about as many books have been written about Roswell as have been written about the Kennedys. In 1991, yet another book, *U.F.O. Crash at Roswell*, written by Kevin Ranole and Donald Schmitt, took things a giant step further: they claimed that in addition to a spaceship, the bodies of several dead aliens had been recovered at the Roswell crash site; the saucer and the bodies were whisked off to Wright Patterson Air Force base in Ohio, before being stored and studied at Area 51 in Nevada.

The claims of a monumental government cover-up are undermined by secret documents released under the Freedom of Information Act and discovered by UFO debunker Philip Klass. The documents clearly show that military intelligence officers were still seeking evidence that UFOs were real well after the Roswell crash. For example, according to the minutes of a March 1948 meeting of the Air Force Scientific Advisory Board, Colonel Howard McCoy, chief of intelligence at Wright Patterson Air Force Base, says: "We are running down every [UFO] report. I can't even tell you how much we would give to have one of these crash in an area so that we could recover whatever they are." If there were an alien spacecraft and alien bodies, and if they were sent to Wright-Patterson shortly after the crash, isn't it likely that the chief of intelligence would have known about it?

In 1994, in response to pressure from a New Mexico congressman, the air force launched its own investigation into the Roswell incident. The final report identified the Roswell wreckage as the remnants of a balloon that was launched as part of Project Mogul, a top-secret project designed to keep tabs on Soviet nuclear tests by carrying acoustical equipment to high altitudes. Special reflectors, constructed from tin foil, wooden sticks, and tape, were attached to the balloons so that they could be tracked by radar. This explains the materials found at the crash site. The Project Mogul records indicate that one such balloon was released on June 4 and by mid-June had drifted to within 20 miles of the Foster ranch where radar contact was lost. The report also stated that "there was no indication in official records from the period that there was heightened military operational or security activity which should have been generated if this was, in fact, the first recovery of materials and/or persons from another world." A Roswell report released in 1995 by the General Accounting Office reached a similar conclusion. Of course, true believers dismiss these reports as part of the cover-up. However, former CIA and Pentagon official-turned UFO investigator Karl Pflock, well-respected in the UFO community, reached the very same conclusion after an extensive two-year investigation.

But what about the alien bodies? Initial 1947 Roswell reports did not mention aliens, and the suggestion that alien bodies were recovered didn't come about until 1991. Witnesses who claim to have seen the bodies were trying to recall events that had happened decades in the past. A second air force report addressed the alien body question and explained that for a few years after 1947, the air force dropped dummies from high-altitude balloons to study the effects of the impact on the human body. The dummies, three and a half to four feet tall with no ears or body hair, closely matched descriptions of the supposed alien bodies. It is possible that the witnesses saw these dummies lying around in the desert and, through the fog of memory, interpreted them as aliens. Overzealous UFOlogists have linked the alien body sightings to Roswell.

Visiting Information

Today, Roswell is a thriving city of about 50,000 in southeastern New Mexico. Since the 1980s, the Roswellian economy has been invigorated by the flying saucer craze, and the city has positioned itself as a major tourist attraction. Should a scientifically minded tourist decide to visit here, an attitude adjustment may be in order. You must relax, take a deep breath, check your rationality at the city limits, and soak up the silliness. On my visit here, I must admit that I, as a committed skeptic, was a little embarrassed to be seen in Roswell, lest I be lumped in with the believers. I just kept reminding myself that I would never see these people again and went on my merry way. Merchants along Roswell's Main Street have taken full advantage of the madness and sell an absolutely bewildering array of souvenir items, all with a requisite little green alien plastered on it. Roswell, the quintessential kitschy tourist trap, is definitely the place to come to buy alien-themed T-shirts, sweatshirts, socks, panties, shot-glasses, license plates, buttons, and dolls.

The main attraction in Roswell is the International UFO Museum and Research Center, located in a converted movie theater right on Main Street. (A new, multimillion-dollar museum is in the planning and fund-raising stage.) The irony is that the UFO museum doesn't actually have any UFOs, only a rather disappointing collection of old photographs, documents, and newspaper clippings. Upon entering, you are greeted by a slender grey alien with an enormous head holding a sign inviting you to sign in and enjoy the visit. Admission is free, although a small donation ($2 for adults, $1 for children) is suggested. Audio tour tapes are available, and you can stick a pin in a map of the United States or the world to show everybody where you're from. The exhibits begin with a timeline of the Roswell incident along with a re-creation of the RAAF communication office. Next, the great cover-up is discussed and exposed. Then you'll come to the back of the museum where a large meeting room is used for lectures by visiting UFOlogists. The exhibits along the opposite wall focus on questions such as "Where did they come from?" and "How did they get here?" Other displays deal with Area 51, the Lubbock Lights, and the Roswell Metal Fragment. Most telling is the "Evidence" exhibit, which is distinguished by an astonishing lack thereof. You'll find only a bunch of fuzzy pictures here. A neighboring exhibit asks: "Crop Circles: Phenomenon or Hoax?" (I'm no UFOlogist, but I think I know the answer.) There are quotes from famous people about UFOs, and throughout the museum you can see UFO, space, and alien-related artwork. Your other-worldly walk ends with a display showing the fake alien used in *Roswell: The*

Kathy Burn-Millyard/Shutterstock

The Alien Costume Contest during the UFO Festival in Roswell, New Mexico.

Motion Picture lying on an examining table. Standing nearby is a mannequin dressed in a surgical gown presumably ready to perform an autopsy.

If you decide to visit Roswell, make every effort to come during the annual four-day UFO Festival held during the first week in July. Otherwise, Roswell may disappoint you. The festival hosts all kinds of out-of-this-world events, including lectures and workshops, panel discussions, planetarium shows, documentary film showings, vendors, live music at the fair grounds and on Main Street, dances, a parade, and even theater. The 2006 festival, for example, featured a workshop "The Alien Mind and the Mind of the Abductee," a lecture on "UFO Secrecy and the Murder of Marilyn Monroe," and *Jocelyn's World,* a play about Earth's first interplanetary foreign exchange student. The highlight of your visit just may be the Galactic Costume Contest where earthlings of all ages (and even a few canines) dress up as their favorite alien. Apparently, there's been a rift between the Roswell Chamber of Commerce and the UFO Museum people, who think the carnival-like atmosphere of the UFO festival demeans the seriousness of their museum and research center. Methinks the museum doth protest too much. A glance at the museum's saucer-bedecked façade and cheesy exhibits reveals how seriously the museum should be taken. In 2007, the festival was taken over by the city of Roswell.

Roswell city officials have announced plans to build a UFO-themed amusement park, complete with an indoor roller coaster that will whisk passengers away on a simulated alien abduction. Initially, the park will cover 60 to 80 acres, with room to expand to 150 acres. Cost estimates run into the hundreds of millions of dollars. Will it ever actually be built? Believe . . .

Hungry for some real science after all the pseudoscience? Head on over to the Roswell Museum and Art Center, where you can find a special exhibit on rocket pioneer Robert Goddard who tested rockets in the wide open spaces outside of

> Websites: www.roswellufomuseum.com
> www.roswellufofestival.com

Roswell in the 1930s. The exhibit includes actual rocket engines and parts used by Goddard and a re-creation of his Roswell workshop.

Area 51 (Groom Lake), Nevada

Area 51, officially known as Air Force Flight Test Center, Detachment 3, a remote six-mile-by-ten-mile block of desert about 90 miles north of Las Vegas, is home to a top-secret air base. The designation "Area 51" is probably a holdover from the days it was part of the Nevada Test Site (NTS) where nuclear weapons were tested during the Cold War. The NTS uses the "Area X" naming scheme to designate various parcels of land. Area 51 contains Groom Lake, a flat dry lake bed, ideal terrain for an air base. Beginning with the U-2 spy plane in the mid-1950s, the air base has been used to develop and test military aircraft including the SR-71 Blackbird and the F-117 Stealth Fighter. No one doubts that Area 51 houses secret goings-on, but much controversy surrounds that secret. UFO aficionados and conspiracy theorists insist that hidden somewhere on (or under) Area 51 is an alien spacecraft recovered at Roswell along with one or more alien bodies. Some even claim that the military is trying to reverse-engineer the super-advanced alien spacecraft. This fantasy was perpetuated in the movie *Independence Day*. Another popular conspiracy theory, presented as part of a special on the Fox television network, holds that Area 51 is where the Apollo Moon landings were hoaxed. In a segment on the venerable TV news magazine "60 Minutes," Lesley Stahl made the rather more believable suggestion that Area 51 is an illegal toxic waste dump. Like Roswell, Area 51 has been absorbed into American popular culture and is regularly featured in TV series, movies, books, music, and even computer and video games. In fact, the Las Vegas minor league baseball team is called the "51s," and a little grey alien is part of their logo.

Visiting Information

If you just can't resist a visit to Area 51, you should understand that there's not much you can actually see. The language used on the warning signs has been toned down from the melodramatic "use of deadly force authorized" (happily, deadly force has never actually been used on any tourist) to more civil threats of fines and possible incarceration. The best plan is to go first to the town of Rachel, population 98 (humans), where a clump of mobile homes serves as the unofficial UFO headquarters. The hottest (and only) hang-out in Rachel is the "Little A 'Le'Inn," a diner and gift shop serving decent diner food, including the ever-popular Alien Burger. Here, you can pick up some Area 51 literature, a local map, and get advice from the locals as to the best spots for approaching the fence. Rooms are available for $40 per night, and RV hook-ups are available. Reservations are recommended.

Before heading out on your alien adventure, you should fill your gas tank. You'll be heading out into the middle of nowhere, and opportunities for refueling are few and far between. From Las Vegas, take I-15 North to U.S.

A lonely stretch of the Extraterrestrial Highway in Nevada.

93 North (Great Basin Highway) at Exit 64. Stay on 93 for about 85 miles. Turn left onto Nevada 318. After ½ mile, veer left onto Nevada 375. You are now on a desolate 98-mile stretch of pavement officially designated by the State of Nevada as the "Extraterrestrial Highway!" Rachel is about 43 miles farther down the road. On your way to Rachel, keep an eye out for the infamous black mailbox about 17 miles down Nevada 375; the large wooden box, painted white, is mounted on a white post with rocks around the base. This mailbox marks the dirt road that leads to the best known entrance to Groom Lake.

A trip to Rachel and Area 51 takes an entire day from Las Vegas. Tour companies offer day trips to Area 51 from Las Vegas for around $200 per person. Check the Las Vegas tourist websites for details.

Websites: www.littlealeinn.com; www.rachel-nevada.com; www.ufo-hyway.com; www.ufomind.com.

9
Space Rocks

I would rather be a superb meteor, every atom of me
in magnificent glow, than a sleepy and permanent planet. Jack London

Rocks, in a wide range of sizes from sand-like grains to super-size boulders, are floating and flitting around in interplanetary space. They are called meteoroids. Some meteoroids are ancient bits of detritus left over from the formation of the solar system some 4.6 billion years ago, a few are pieces of larger celestial objects like Mars or the Moon, but most meteoroids are pieces of much larger rocks called asteroids.

Most asteroids inhabit the space between the orbits of Mars and Jupiter, appropriately designated as the asteroid belt, and can have diameters measured in miles. In fact, 26 asteroids have diameters measuring at least 120 miles. By far the largest asteroid, Ceres, has a diameter greater than 500 miles. (In 2006, the International Astronomical Union reclassified Ceres as a "dwarf planet.") The asteroids are thought to be objects that weren't quite able to coalesce into a planet, probably because of the gravitational pushing and pulling of the local bully, the planet Jupiter. Sometimes, an asteroid can be gravitationally prodded out of its comfortable orbit and into a more elongated, eccentric orbit that intersects the orbit of the Earth. On thankfully rare occasions, a large "Near Earth Asteroid" is placed on a collision course with the Earth. Such a colossal collision 65 million years ago is thought to have brought about the sudden extinction of the dinosaurs. This will inevitably happen again. An effort is currently under way to identify all large Near Earth Asteroids and to calculate their orbits. By using current technology an asteroid on a collision course with the Earth can be nudged off course so as to narrowly miss our home planet, but all the contingency plans depend upon finding the asteroid in plenty of time to enact the countermeasures.

Meteoroids sometimes hit the Earth's atmosphere where friction between the air and the rock cause it to heat up and glow. Such a glowing space rock is called a meteor or "shooting star," and we see it as a fleeting streak of light across the dark night sky. On a clear, moonless night, several meteors can usually be seen every hour. A few times each year, the orbit of the Earth intersects the orbit of a comet, which has left a trail of rocky debris in its wake. This results in a dramatic increase in the number of meteors, and observers on Earth witness a meteor shower.

The vast majority of meteors are so small that after zipping through the atmosphere only microscopic specks of ash are left to gently rain down onto the Earth. A few meteors are large enough to survive the trip and actually hit the surface of the Earth. A rock that has fallen from space and crashed on the Earth's surface is called a meteorite, which is usually named after the place where it landed. Scientists estimate that somewhere around a hundred tons of meteoric material, mostly in the form of dust and ash, falls to the Earth every week!

Meteors can be categorized into three basic types: stony, iron, and stony-iron. Stony-iron meteorites, the rarest type, consist of roughly equal quantities of rock and metallic material. Iron meteorites are composed of iron along with the metal nickel. Although they form a relatively rare type of meteorite, they are common in museum collections because they are not only easy to distinguish from Earth rocks but also weather resistant, and therefore easy to find. Iron meteorites are believed to be chunks of metal from the cores of asteroids.

More than 95 percent of all meteorites are the stony type composed mostly of silicate minerals but also containing grains of metallic iron. Stony meteorites can be further subdivided into chondrites and achondrites. Chondrites, the most common type of stony meteorite, feature small spherical structures about a millimeter in diameter called *chondrules* (Greek for "small sphere"). The much less common achondrites are missing these little spheres. Of particular interest is a rare type of chondrite called a carbonaceous chondrite. As the name implies, these meteors contain high levels of water and carbon-based compounds, such as amino acids. These meteorites contain complex organic compounds, which suggests that the building blocks of life are easy to make and can be found in abundance throughout the universe. This fact has implications for the possibility of extraterrestrial life. A few scientists have even gone so far as to suggest that life on Earth is the result of "seeding" from these types of meteorites.

Usually, the Earth's atmosphere adequately defends the surface against invading space rocks. But when a sufficiently large asteroid or meteor pays us a visit, the Earth is utterly defenseless. Friction with the Earth's atmosphere doesn't slow these behemoths down very much, and they slam into the Earth at high speeds punching a circular hole, a crater, in the surface. When the rock hits, the prodigious energy it possesses due to its heavy mass and high speed is transformed into heat in a fraction of a second. As a result, much of the meteor and some of the rock around the point of impact are melted or vaporized. Shock waves blast the surface rocks upward and radially outward. Upturned strata and a circular rim raised above the surrounding terrain are tell-tale signs of an impact crater.

An abundance of impact craters are evident on Mercury, Mars, and many of the moons of the outer planets. When we look at our own Moon, we see that it is pockmarked with thousands of craters. But the Earth and Moon are next-door-neighbors in space, so if the Moon has craters, why not the Earth? The difference is that the Moon is, geologically speaking, a dead world. Except for an occasional new impact crater, the surface of the Moon is fixed and frozen in time. The Earth, however, is geologically active. The surface of the Earth is constantly changing through earthquakes, erupting volcanoes, drifting continents, and the erosive action of the wind and rain. Over millions and billions of years, these geological forces erase any vestiges of cratering.

For the first half of the twentieth century, a heated debate raged among astronomers regarding the origin of the Moon's craters. For a long time, astronomers had just assumed that lunar craters were volcanic. By 1900, a few astronomers and geologists had begun to question this assumption and advanced the notion that they may be impact craters. A major objection to this hypothesis was that all the lunar craters are perfectly round. Some astronomers argued that most meteorites would hit the Moon's surface at an angle, producing elongated, oval-shaped craters rather than circular craters. Mining engineer Daniel Barringer put this objection to the test by firing rifle bullets into rocks and mud. He discovered that even a projectile impacting a surface at an angle would still leave a circular hole, just as it would if it hit the surface directly. In 1923, Barringer published his findings in an article in *Scientific American*. In spite of this, not until the 1960s did scientists finally become absolutely convinced that the craters on the Moon were impact craters and that a few impact craters exist here on Earth.

The first meteor impact crater to be identified was Meteor Crater located in the Arizona desert. Today, as a result of aerial photography, satellite imaging, and a better understanding of the geological signs of impact craters, about 200 impact sites around the world have been identified, both on land and under the ocean. The craters range in diameter from a few dozen yards to more than 60 miles across and range in age from a few thousand to more than two billion years. The vast majority of these craters are barely recognizable, while a precious few are clear, sharp, and well-defined. In the United States, two well-defined impact craters are easily accessible to the public: the famous Meteor Crater and the not-so-famous Odessa Crater in Texas.

Aside from meteors, the other kind of space rock that can be viewed here on Earth are the rocks returned from the Moon by the Apollo astronauts. The six successful Apollo lunar landing missions returned a total of 842 pounds of rock and soil from various areas of the Moon, most of which are stored in the Lunar Sample Building at the Johnson Space Center in Houston, Texas. To insure that some Moon rocks would survive in case a major disaster were to strike the Johnson Space Center, a smaller collection is held at the Brooks Air Force Base in San Antonio, Texas. A few samples were distributed to select museums across the country. The Moon rocks, a national treasure, are considered priceless. In 1993, three tiny fragments, that weighed only 0.2 grams and had been collected by the unmanned Russian Luna 16 mission, sold for nearly $450,000.

The three most common types of Moon rocks are breccias, basalts, and various types of feldspar. Breccias are rocks consisting of an amalgamation of bits and pieces of other rocks. They formed as a result of meteor impacts that smashed rocks into fragments that were then welded together by the heat generated by the impact. Basalts are dense, dark-colored rocks formed when molten magma cools and solidifies. When you look at the Moon in the night sky, the dark areas, called *maria* (Latin for "seas" because early astronomers thought they were water), are made mainly of basalt. The lighter areas on the Moon's surface are made of low-density feldspar, a common mineral found on Earth. Lunar "soil" isn't at all like soil we have here on Earth. There is, of course, no organic material in lunar soil; it's just tiny pieces of rock that have been pulverized by the meteor impacts. A number of new minerals were discovered on the Moon including a mineral called Armalcolite in honor of the three Apollo 11 astronauts (*Arm*strong, *Al*drin, and *Col*lins). At the time of the Apollo missions, Armalcolite was not known to exist on Earth, but it has since been found to exist near impact craters.

Moon rocks and Earth rocks are very similar, but, because there is no water on the Moon (except for some water-ice in shaded craters at the poles), there are no sedimentary rocks such as limestone or sandstone which depend on the action of water. Also, due to a lack of erosion or plate tectonics, the Moon rocks are much older than rocks on Earth. The rocks brought back from the Moon range in age from about 3.2 billion years old to about 4.5 billion years old, whereas the very oldest Earth rocks are a mere 3.8 billion years old.

The most famous space rock in the world has got to be ALH 84001. This meteor, the first meteor (001) found in the Allan Hills (ALH) region of Antarctica in 1984, hails from Mars. So how did it end up here on Earth? Millions of years ago, a large rock from space hit the surface of Mars and knocked this four-pound chunk of Mars out into interplanetary space. After floating around in space for millions of years, the rock was grabbed by the gravitational grip of the Earth and crashed into Antarctica about 13,000 years ago. After being identified as a possible Martian meteorite, the rock ended up at the Johnson Space Center in Houston for study and analysis by NASA scientists. How do we know it's from Mars? A chemical analysis of tiny pockets of gas trapped inside the rock reveals that the chemical composition of the gas exactly matches that of the atmosphere of Mars. No one disputes that fact; this is definitely a rock from Mars. The rock became world renowned in 1996 when a team of NASA scientists announced that ALH 84001 contained evidence that microbial life once existed on Mars. The NASA team claimed that the best explanation for the presence of certain minerals, molecules, and structures found within the rock was that they were formed by living organisms. In fact, the NASA scientists identified certain structures as actual microfossils of organisms that had once been alive on the surface of Mars. The announcement created a media sensation but was met with skepticism in the scientific community. Other scientists came up with alternative nonbiological explanations for the structures. Of course, no one knows for sure. The only way to conclusively answer the question of life on Mars is to go to Mars and look. In any case, this little rock has renewed interest in Mars and as a result, over the last decade, NASA has sent a number of probes to Mars in search of answers. A piece of ALH 84001 can be seen at the National Museum of Natural History in Washington, D.C.

The first two sites described below are the best preserved meteor craters in the United States. Of the two, the Meteor Crater in Arizona is definitely the one to see. The Odessa crater is barely recognizable as a meteor crater. The final five sites are some of the best museum and university collections of meteorites and Moon rocks. These sites are arranged according to the size of the collection.

Meteor Crater, Arizona

This is the most famous, best preserved, and most extensively studied meteorite impact crater in the world. It is not only the first terrestrial crater proven to be an impact crater, but it is also the closest we earthlings can come to experiencing the cratered terrain of the Moon, Mars, or Mercury. A photo of this crater can be found in nearly every astronomy textbook. The crater was formed about 50,000 years ago when an iron-nickel meteorite (now known as the Canyon Diablo meteorite) about half the width of a football field and weighing several hundred thousand tons slammed into the Earth. Traveling at an estimated speed of 28,600 mph, the explosive

Mary Lane/Shutterstock

Meteor Crater in Arizona.

impact released an amount of energy equivalent to 2.5 million tons of TNT or about 150 times the energy of the Hiroshima atomic bomb. The collision displaced 175 million tons of rock and produced a crater more than 4,000 feet across and about 570 feet deep surrounded by a rim of rock rising 150 feet above the surrounding terrain. Although it may seem obvious from a glance at the photograph that this is a meteorite impact crater, it took the scientific community until the 1960s to reach a consensus as to the origin of the crater.

The first serious scientific investigation of the crater was undertaken in 1891 by Grove Karl Gilbert, chief geologist for the U.S. Geological Survey and the most respected geologist of his generation. Gilbert first considered the possibility that this was a volcanic crater, but he eliminated this explanation because no volcanic rock was found at the site. Gilbert then considered the two remaining hypotheses: one was that it was created by an explosion of steam caused by volcanic activity deep within the Earth; the second possibility was that it was an impact crater. The second hypothesis was supported by the fact that fragments of iron and nickel had been found on the land surrounding the crater.

Gilbert devised two crucial tests to decide between these competing hypotheses; he based his tests on the incorrect assumption that the meteorite had survived the impact and was buried beneath the crater floor. The first test was to carefully observe the behavior of magnets and compass needles in and near the crater. A large mass of buried iron should have an affect on their behavior. The second test involved estimating the total volume of ejected rock forming the crater rim. If the crater had been formed by a volcanic steam explosion, then one would expect that the volume of ejected material would be equal to the volume of the empty crater. If, however, the crater had been formed by a meteorite impact, then one would expect the volume of the ejected material to be greater than the volume of the empty crater because the buried meteorite would take up a significant amount of space below the crater floor.

The meteor hypothesis failed both tests. A variety of experiments showed that the crater had no affect on the behavior of magnets or compasses, and, according to Gilbert's calculations, the volume of the rim material was just equal to the volume of the crater. Thus, the only surviving hypothesis was the steam explosion idea, and, because Gilbert's word was geological gospel, nobody dared to question his conclusion. The presence of meteoric material near the crater was, according to Gilbert, coincidental.

In 1902, Daniel Moreau Barringer, a well-known and successful mining engineer, heard about the crater in a casual conversation with a friend. A few months later, he learned that small spheres of meteoric iron were mixed in with the ejected rocks in the crater rim. This tell-tale fact led Barringer to the immediate conclusion that the formation of the crater and the impact of a meteor must have happened at the same time. Otherwise, the ejected rock and the meteoric material would have been found in separate layers. He also assumed, as Gilbert had, that the meteorite would have to be about as large as the crater itself and must lie intact and buried under the crater floor. Barringer thought he could mine the iron from the crater and make a fortune, so, without ever having seen the crater, he formed the Standard Iron Company and began securing the mining rights to the crater.

For the next 27 years, Barringer and his financial partners spent, in today's money, more than ten million dollars trying to recover iron from the crater. The mine produced no iron and failed financially, but it proved to be a scientific success; in the process Barringer gathered evidence that eventually convinced most of the scientific community that the crater was, in fact, formed by a meteorite impact. In 1906, Barringer, along with a physicist colleague, presented his evidence for a meteorite impact to a skeptical audience at the Academy of Natural Sciences in Philadelphia.

The tide in the debate began to turn somewhat when the eminent geologist George Perkins Merrill visited the crater in 1907 and published a series of papers supporting Barringer's conclusions. Merrill based his findings on an analysis of a type of quartz glass found at the crater that could only be produced by extreme heat. This glass is very similar to the fulgurite glass that is sometimes formed when lightning strikes sand. Merrill also argued that because the rock layers below the crater were undisturbed, the force that created the crater could not have come from below, as in a steam explosion, but rather from above, as in a meteorite impact.

Nevertheless, the crater controversy continued. Part of the geologists' reluctance to embrace Barringer's explanation may have stemmed from the prevailing principle of "uniformitarianism," the idea that geological processes happen gradually over millions of years and that scientists should avoid sudden, catastrophic explanations for geological phenomena. Another factor that probably contributed to this resistance was Barringer's forceful and, at times, abrasive personality, a flaw that did not endear him to the geological community. For example, Barringer began his 1906 paper with a tactless attack on Gilbert; Barringer said that he could not understand how any experienced geologist could have failed to recognize the evidence showing that the crater could not have possibly resulted from volcanic forces.

Although more and more scientists gradually became convinced that Barringer was right about the origin of the crater, they also reached a secondary conclusion that devastated Barringer. Whereas Barringer had always assumed that the meteorite had survived the impact, prominent astronomers, including Harlow Shapley and Henry Norris Russell, began to realize that the meteorite most probably totally vaporized upon impact. Convinced by the weight of scientific opinion that there was no iron to recover, the board of directors of the Meteor Crater Exploration and Mining Company voted to halt all work at the crater in September 1929. In late November of the same year, Daniel Barringer died of a massive heart attack.

In 1946, the legendary meteorite expert Harvey Ninninger performed a careful analysis of the soil surrounding the crater and found tiny metal particles and small "bombs" of melted rock mixed in with the soil. He concluded that both kinds of particles condensed out of a cloud of vaporized metal and rock that formed from the enormous heat produced by the impact. Recent computer simulations of the event prove that most of the meteorite did, in fact, vaporize on impact and was transformed into a fine mist of molten metal that rained down on the surrounding desert.

The final confirmation that the Barringer crater was created by a meteorite that exploded upon impact came in 1963 when Eugene Shoemaker compared the Barringer crater to the craters formed by underground tests of nuclear weapons in Nevada. After mapping out the order of the layers in the underlying rock in the bomb craters and comparing it to the order of the layers in the ejected rock, he found that the ejected rocks were deposited in the reverse order, just like at the Barringer crater. In fact, Shoemaker showed that the nuclear craters and the Barringer crater were structurally similar in every respect. Another key discovery was the presence of coesite and stichovite in the crater. The only known mechanism for producing these rare minerals is by hitting quartz containing rocks with a tremendous force or shock wave. Shoemaker's discovery was the final, definitive proof that an extraterrestrial object had actually hit the surface of the Earth.

Today, the crater is still owned by the Barringer family and is finally earning them some well-deserved cash—not as an iron mine, but as a popular tourist attraction. Scientific research is still conducted at the crater, and the Apollo astronauts trained there for the lunar landings.

Visiting Information

Meteor Crater is located off of Interstate 40 at Exit 233, 35 miles east of Flagstaff and 20 miles west of Winslow, Arizona. From Memorial Day through September 15, the Visitor Center is open from 7:00 A.M. until 7:00 P.M. daily. The hours are 8:00 A.M. through 5:00 P.M. the rest of the year. The center is closed on Christmas. The admission fee is $15 for adults, $13 for senior citizens 60 and over, and $6 for children ages 6 to 17.

The Meteor Crater Visitor Center is home to an Interactive Learning Center featuring two dozen exhibits focusing on meteor impacts and collisions. Among the exhibits are the Holsinger Meteorite, which, at 1,406 pounds, is the largest fragment of the Canyon Diablo meteorite ever found; photographs of craters on Mercury, Venus, Mars, and the Moon taken from NASA spacecraft; images of the Shoemaker-Levy comet crashing into Jupiter in 1994; a Crater Location Spotter that identifies points of interest and soil types within the crater; and "Crater Perspectives" where you can hear about the exploits and discoveries of Gilbert, Barringer, and Shoemaker. The "Create-A-Crater" exhibit allows you to select the diameter, velocity, and angle of descent of a meteor and see what the resulting crater looks like. A ten-minute video, "Collisions and Impacts," is shown twice every hour in the theater. Behind the Learning Center four observation areas provide views of the crater. Tours are available to take you for a walk of about a third of a mile along the crater rim. These one-hour tours leave every hour starting at 9:15 A.M., with the last tour leaving at 2:15 P.M. Visitors are not allowed to hike down into the crater.

Back at the Visitor Center, you will want to browse through the gift shop that includes an interesting rock shop. There is a Subway restaurant where you can grab a bite to eat and an Astronaut Memorial Park where you can relax and maybe enjoy a pic-

Website: www.meteorcrater.com
Telephone: 1–800–289–5898

nic. There are no accommodations at the crater, but the Meteor Crater RV Park is about five miles from the crater.

Odessa Meteor Crater, Odessa, Texas

The Odessa Meteor Crater was formed about 50,000 years ago when an iron meteorite weighing an estimated one thousand tons crashed into the Earth. The impact displaced 4.3 million cubic feet of rock and formed a crater more than 500 feet wide and about 100 feet deep. Over the millennia, the wind and rain deposited silt into the crater so that today, the crater floor is a mere five to six feet below the level of the surrounding terrain. Four smaller craters have been discovered near the main crater, all of which have been completely filled in with dirt. These smaller craters, ranging in size from 15 to 50 feet wide and from 7 to 17 feet deep, formed from the impact of several smaller meteorites that accompanied the main meteorite. The geological structure of the main crater is significantly different from the structure of the Barringer meteor crater in Arizona. Today, most impact craters in the world are classified as either "Barringer Type" or "Odessa Type," depending on the geology.

The Odessa Crater was first identified as a meteor impact crater in 1926 when mining engineer Daniel Barringer, who had learned of a possible meteor crater in Texas from a letter published in a mining journal, sent his son Daniel Barringer, Jr., to Texas to have a look. After a careful investigation eliminated all other possibilities, Barringer Jr. confirmed that the crater was formed by a meteorite and, using a prearranged code, telegraphed his conclusion to his father. Thus, the Odessa Crater became only the second meteor crater to be identified in the United States (the first, of course, was Meteor Crater in Arizona).

During the Great Depression, the WPA drilled a 165-foot hole into the middle of the crater in an unsuccessful attempt to find the meteor. Later attempts also proved fruitless, leading scientists to conclude that the meteor must have vaporized upon impact. Nevertheless, several tons of meteoric material have been recovered from the crater with the single largest fragment weighing in at 300 pounds. Through the years, the crater has been marred by random and careless excavation. Upon visiting the crater, meteor expert Harlow Ninninger referred to the overzealous digging as "Odessacration."

Visiting Information

The Odessa Meteor Crater, which has been designated as a National Natural Landmark, is located approximately four miles west of Odessa, Texas. To get there from Odessa, drive west on I-20 and take Exit 108. Drive south two miles on Meteor Crater Road and follow the signs. A Visitor Center houses a small museum with chunks of the actual meteorite that created the crater accompanied by a few meteor-related displays. A short well-marked trail takes you through the crater and up to the west rim and back. The trail has interpretive signs describing the various features of the crater. The Visitor

Center is open Tuesday through Saturday from 10:00 A.M. to 5:00 P.M. and on Sunday from 1:00 P.M. to 5:00 P.M. Admission is free. A word of warning: don't come here expecting to see a well-defined, spectacular meteor crater similar to the one in Arizona; if you do,

> Website: www.netwest.com/virtdomains/ meteorcrater/About.htm
>
> Telephone: 915–381–0946

you will be extremely disappointed. But, if you need something else scientific to see in Odessa, you might visit the replica of Stonehenge on the campus of the University of Texas branch.

National Museum of Natural History, Moon, Meteorites, and Solar System Gallery, Washington, D.C.

With more than 17,000 specimens from more than 9,000 distinct meteorites of every type, the National Meteorite Collection of the Smithsonian Institution is one of the largest and best meteorite collections in the world. The collection includes nearly 7,000 polished wafer-thin cross-sections of meteoric rock mounted on glass. These thin sections are made available to scientists who are interested in studying the structure, texture, and composition of meteorites. Most notably, the collection contains fragments of seven of the thirteen known Martian meteorites. Only a fraction of the collection is on public display in the Moon, Meteorites, and Solar System Gallery of the National Museum of Natural History.

The highlight of the collection here is undoubtedly a piece of the Alan Hills meteorite from Mars (ALH84001) that was described in the introduction to this chapter. The meteorite is sitting rather inconspicuously at the end of a shelf in a display case. No special signs draw your attention to this treasure, so you'll have to hunt for it. Across the gallery, another well-known Martian meteorite, the Nakhla meteorite, is displayed openly so that you can reach out and, as the sign suggests, "touch a piece of Mars." The Nakhla meteorite fell in a region of Egypt called Nakhla near Alexandria in 1911. The meteor broke into 40 pieces with a combined weight of about 20 pounds. A farmer claimed that a piece of the meteor actually hit and vaporized his dog, but the story is considered, at best, apocryphal. The Nakhla meteorite is black inside with orange veinlets running throughout. Scientists think the meteorite crystallized out of Martian lava about 1.3 billion years ago. In 1999, the same team of NASA scientists that examined the Allan Hills meteor looked at Nakhla and claimed that it, too, showed signs of fossilized Martian bacteria. One argument that can always be made against the extra-terrestrial origin of carbon-containing material found on a meteorite is that the surface was contaminated when it landed on Earth. In 2006, a piece of the Nakhla meteorite from the collection at London's Natural History Museum was broken open so that fresh samples that had never been exposed to the Earth's environment could be obtained. A carbon-rich mate-rial was found filling tiny cracks within the Nakhla meteorite. The material closely resembles features associated with microbial activity in volcanic glass from the Earth's ocean floor.

In addition to meteorites, the gallery has the best display of Moon rocks I have seen—much better than the display at the National Air and Space Museum. Most Moon rocks on public display are quite small, but this exhibit has a half-dozen or so large rocks mounted upright on metal clamps. The Moon rocks on display here, fragments of larger rocks collected by the Apollo astronauts, were selected to represent the basic types and ages of rock. Each rock is accompanied by an informative plaque giving the age and type of rock.

Another fascinating exhibit in the Meteorite Gallery addresses the ques-tion of what caused the sudden extinction of the dinosaurs 65 million years ago. The leading hypothesis, first proposed in 1980 by Louis and Walter Alvarez, is that a giant asteroid about six miles in diameter smashed into the Earth with a force equal to 300 million atomic bomb blasts. The impact spewed dust and smoke into the atmosphere that blocked much of the Sun's light. The Earth was plunged into a dark, cold period lasting many months. After the dust settled, greenhouse gases that had been released by the impact may have caused global temperatures to skyrocket. These rapid and extreme climate changes resulted in the extinction of not only the dinosaurs but also as much as 70 percent of all plant and animal species. What evidence sup-ports such a catastrophic impact? The metal iridium, quite rare on the sur-face of the Earth, is found in abundance in asteroids. Interestingly enough, all over the world, very high concentrations of iridium are found in the rock layer known as the Cretaceous-Tertiary boundary (sometimes abbreviated as the K-T boundary). The age of this rock layer is . . . drum roll, please . . . 65 million years! Here in this exhibit, you can actually see a large sample of rock showing the iridium rich K-T boundary. But if a large asteroid hit the Earth, wouldn't it leave a crater? Yes, and in fact, a 90-mile-wide ancient crater has been discovered off the coast of the Yucatan peninsula in Mexico.

This crater, coupled with the iridium layer found at the K-T boundary, is strong evidence in favor of an asteroid impact.

Visiting Information

The National Museum of Natural History is part of the Smithsonian Institution on the National Mall in Washington, D.C. Admis-

Website: www.mnh.si.edu/
Telephone: 202–633–1000

sion is free. Regular hours are 10:30 A.M. to 5:30 P.M. daily except Christmas. Check the website for extended hours in the summer.

American Museum of Natural History, Arthur Ross Hall of Meteorites, New York City

The Arthur Ross Hall of Meteorites is a single large circular room within this cavernous museum with meteor-related displays arranged mostly around the circumference. Upon entering the hall your attention is immediately drawn to the center of the room where you'll see the Ahnighito meteorite, a huge fragment of the Cape York meteorite. Weighing in at a hefty 34 tons, this iron meteorite is the largest meteorite on display in a museum anywhere in the world. Only two meteorites are larger than this: the Hoba meteorite in Namibia at 60 metric tons and the Chaco meteorite in Argentina at 37 metric tons; both of these meteorites remain at or near the impact point. The Ahnigito meteorite rests on a cluster of six supports that extend through the bottom of the building and are anchored in bedrock. The Cape York meteorite was discovered in 1894 by Arctic explorer Robert Edwin Peary in the Melville Bay region of northwestern Greenland. The museum website has a film showing photographs of the recovery of this giant meteorite. Nearby are two smaller fragments of the Cape York meteorite, nicknamed the "woman" and the "dog." Legend has it that the meteorites were once a sewing woman and her dog who were cast out of heaven by an evil spirit. Ahnighito was the tent they shared. Some historians have suggested that this fanciful tale was invented for Peary's benefit.

The hall displays fragments from many of the best-known meteorites, including the Hoba and Wolf Creek meteorites and the Orgueil, Murchison, and Allende carbonaceous chondrites. The hall also features fragments of four Martian meteorites, including a piece of the Nahkla meteorite discussed in the previous entry.

A fragment of the Peekskill, New York, meteorite is on display. The flight of this meteor was caught on videotape by fans at a high school football game in 1992. A bowling-ball sized fragment of the meteorite smashed the trunk of a red 1980 Chevy Malibu parked in the owner's driveway—one of the rare cases where a falling meteor resulted in property damage. Actually, the encounter with the meteorite increased the value of the car. Before it was hit by the meteorite, the car was worth about $500; after the impact, it sold for $10,000!

Several exhibits deal with the Canyon Diablo meteorite, the space rock responsible for the Meteor Crater in Arizona. A diorama of the meteor crater is presented along with diagrams describing its geological structure. Several fragments of the actual meteorite are on display, including the second largest fragment ever found.

By far the most beautiful meteorites are on display in the "Jewels from Space" exhibit. These rare meteorites called pallasites, represent only about 1 percent of known meteorites. Pallasites are a type of iron meteorite that contains jewel-like olivine crystals. They are thought to originate from the core-mantle boundary of large asteroids.

It is easy to miss the three large Moon rocks displayed vertically one above the other. The rocks are surrounded by interesting information about the Moon, but there is no large sign exclaiming "Moon Rocks"! The top Moon rock is a type of breccia called anorthosite, rich in the mineral feldspar. This rock was collected from the Descartes highland region by the *Apollo 16* astronauts in April 1972. The middle rock is a mare basalt or dark lava collected at the eastern edge of Mare Serenitatis (the Sea of Serenity) by the *Apollo 17* astronauts in December of 1972. The bottom rock is another type of basalt called KREEP. The letters of the acronym stand for the chemical elements found in the rock: "K" is for potassium (potassium is represented by the letter K on the periodic table), "REE" is for rare Earth elements, and "P" is for phosphorus. This rock was collected by the *Apollo 14* astronauts in the Fra Mauro highlands in February 1971.

Visiting Information

The American Museum of Natural History, one of the top science museums in the world, is located at Seventy-ninth Street and Central Park West in New York City. The main entrance to the museum is on Central Park West, but the easiest way to get to the museum is to take the B or C subway line to Eighty-first Street. There is an entrance to the museum at the subway stop, and that ticket line is usually shorter than the lines at the main entrance. The museum is open every day from 10:00 A.M. to 5:45 P.M. except Thanksgiving and Christmas. On the first Friday of every month, the museum stays open until 8:45 P.M. and you can listen to live jazz in the evening. General Admission to the museum is $14 for adults, $8 for children ages 2 to 12, and $10.50 for senior citizens and students with identification. General admission does not include a planetarium show or a movie. For those, you'll have to buy a higher-priced combination ticket. You can avoid ticket lines altogether by purchasing your tickets online. You will no doubt want to combine your visit to the Hall of Meteorites with a visit the museum's Rose Center for Earth and Space (included in your general admission). The Rose Center is described in detail in chapter 4 on planetaria.

Dining options include the Museum Food Court, which is directly accessible from the lower level of the Rose Center; the Café

> Website: www.amnh.org
> Telephone: 212–769–5100

Pho, which serves Vietnamese cuisine in the Seventy-seventh Street lobby; and the Café On 4, serving light fare while providing a view of the museum grounds.

Center for Meteorite Studies, Arizona State University, Tempe, Arizona

With fragments from more than 1,555 separate meteorite falls, Arizona State University's Center for Meteorite Studies is home to the world's largest university-based meteorite collection and is the fourth largest meteorite collection in the world. A small part of the collection is on public display in the meteorite museum. If you think you've found a meteorite, the center will examine your specimen for free. They will return your specimen if it is not a meteorite; if it is, the center will offer to buy it.

Visiting Information

The museum is a single room in the Bateman Physical Sciences Center at Palm Walk and Univer-

> Website: http://meteorites.asu.edu
> Telephone: 480–965–6511

sity Drive on the campus of Arizona State University in Tempe, Arizona. The museum is open Monday through Friday from 9 A.M. to 5 P.M., and admission is free.

Oscar E. Monnig Meteorite Gallery, Texas Christian University, Fort Worth, Texas

Oscar Monnig was a Fort Worth business man with a passion for astronomy and meteorites. As a young man, he traveled to the major natural history museums across the country to view meteorites but returned from the trip discouraged because the curators had treated him "as a nobody" and did not offer to show him the meteorites. In the early 1930s, Monnig decided to start his own collection. He formed a wide network of people, including local newspaper editors, who alerted him to possible meteorite falls. He offered to buy meteorites from Depression-era "dustbowl" farmers at the princely rate of a dollar per pound. For the lucky farmers who found a meteorite, Monnig's check was sometimes the most money the farm produced that year. His collection eventually grew to approximately 3,000 specimens from 400 different meteorites, making it one of the largest private collections in the country. Late in his life, Monnig donated his meteorites to Texas Christian University to ensure that the collection would remain in the city that had supported his family business. About a hundred of the best specimens are on display in this gallery. Notable meteorites that can be seen here include fragments of the Odessa and Peekskill meteorites, the nickel-rich Tishomingo meteorite, and the unusual carbonaceous chondrite Bells.

Visiting Information

The Monnig Gallery is located on the campus of Texas Christian University in the Sid Richardson Science Building at

Website: www.monnigmuseum.tcu.edu
Telephone: 817–257–6277

the corner of West Bowie Street and Cockrell. The gallery is open from 1:00 P.M. to 4:00 P.M. Tuesday through Friday and from 9:00 A.M. to 4:00 P.M. on Saturdays. The gallery is closed on university holidays. Audio wands are available. Admission is free.

Meteorite Museum, University of New Mexico, Albuquerque, New Mexico

The University of New Mexico's Institute for Meteoritics owns a meteorite collection that includes specimens from more than 600 different meteorites, making it one of the largest collections in the United States. Part of the collection is on public display in the Meteorite Museum. Here, you can see samples of all the main types of meteorites, including Norton County (the second largest stony meteorite in the world), fragments from the meteor that formed Meteor Crater in Arizona, a meteorite that holds within it the oldest known rock particles in the solar system, a piece of a Martian meteorite, and a 1,600-pound fragment of the Navajo iron meteorite. Also on display are meteorites that have been found in New Mexico and samples of various kinds of rocks that are created in the formation of an impact crater.

Visiting Information

The Meteorite Museum is a single room located in Northrup Hall on the campus of the University of New Mexico in Albuquerque. The museum is open Monday through Friday from 9:00 A.M. to 4:00 P.M. and admission is free. You can pick up a brochure in the museum for self-guided tours.

Website: epswww.unm.edu/iom
Telephone: 505–277–1644

10
Top Ten
Out-of-This-World
Experiences

*Twenty years from now, you will be more disappointed by the things
you didn't do than by the ones you did do. So throw off the bow lines.
Sail away from the safe harbor. Catch the trade winds in your sails.
Explore. Dream. Discover.*

Mark Twain

In this final chapter, I'm going to do something a little different. Rather
than list space-related sites, I'm going to describe some space-related
experiences. All of them, except #10, involve travel, and at least one may be
beyond the budget of most scientific travelers. It seems like everybody has a
"Top 10" list these days, so I'll follow suit. Of course, there are lots of other
"spacey" things you can do. Some universities and museums have space or
astronomy camps for kids. Your local college or university may let you sit in
on an astronomy class. Amateur astronomy clubs can be found nearly every-
where. You might even want to invest in a small telescope; you can get a
good one for a few hundred dollars. Or just get a map of the night sky, drive
way out of town on a clear night to escape the lights, and see how many
stars, planets, and constellations you can identify. Ready? Let's start the
countdown . . .

10. Help Search for E.T. with SETI@home

You can actually take part in the search for life on other worlds by volunteering your computer for the SETI@home project. SETI is an acronym for the Search for Extraterrestrial Intelligence, the scientific quest to find intelligent life beyond the Earth. One method is to sit back and listen for a signal of intelligent origin in the radio part of the electromagnetic spectrum. But there are lots of frequencies to listen to and lots of sky to look at and consequently lots of data to analyze. This requires a tremendous amount of computing power. Normally, radio SETI projects use supercomputers located at the observatory to perform most analysis. Then computer guru David Gedye got the bright idea of creating a sort of virtual supercomputer by combining the power of a large number of Internet-connected computers. Individuals can participate by running a free program called SETI@home that downloads and analyzes radio telescope data. The program uses part of your computer's CPU power, disk space, and network bandwidth, but you can control exactly how much of your computer's resources are used and when. The observational data is collected by the Arecibo Radio Telescope in Puerto Rico and recorded on tapes that hold more than fifteen hours of observations. The tapes are then mailed to the project's headquarters at the University of California at Berkeley where it is divided into "work units" consisting of 107 seconds of data. The work units are then sent out over the Internet to people all over the world for analysis. The SETI@home project has well over a million computers in the system and has been acknowledged by the *Guinness Book of World Records* as the largest computation in history. Since its launch in 1999, the project has logged over two million years of total computing time.

Unfortunately, the project has not yet found any signals that are unequivocal signs of extraterrestrial intelligence, although several candidate signals have been identified for further analysis. The project plans to use a radio telescope in Australia to expand the search into southern hemisphere skies.

To learn more about the project or to download the program, go to the project's website.

Website: http://setiathome.berkeley.edu

9. Stay at a Telescope Motel or Live in an Astronomy Community

Telescope motels are places where you can go and indulge your astronomical fantasies. Telescopes and dark, clear skies offering good viewing are the main attractions here. Two of the best known telescope motels in the United States are described below along with a lodge at a major observatory that is open to the public.

Star Hill Inn

Situated on 200 acres in Sapello, in northwestern New Mexico, the Star Hill Inn was the nation's first astronomical retreat. At an elevation of 7,200 feet, the dark night skies here offer terrific viewing. The Inn offers eight nicely appointed guest houses ranging in size from one to three bedrooms and ranging in price from $170 to $380 per night with a two-night minimum stay. Each guest house has a kitchen and a fireplace. A variety of high-quality telescopes ranging in size from 7 to 22 inches are available for use by guests at no extra charge, but you must reserve the telescopes in advance to insure availability. A 24-inch telescope is available for an extra charge of $100 per night, and you can rent a CCD camera

Website: www.starhillinn.com
Telephone: 505–429–9998

for $50 per night. During the day, guests may venture to a number of nearby attractions, including Taos and Sante Fe.

The Astronomer's Inn

This bed and breakfast located on a hilltop about a one-hour drive east of Tucson, Arizona, features four themed guest rooms and is home to the privately owned Vega-Bray Observatory. Room rates range from $85 to $185 with breakfast included. The observatory offers a wide range of telescopes from a 3-inch refractor to a 20-inch Maksutov reflector. There is an additional fee for using the telescopes, and you must use them under the guidance of an astronomer who can customize your session based on your knowledge and interest. A four-hour observing session costs

Website: www.astronomersinn.com
Telephone: 520–586–7906

from $95 to $130. Experienced observers can rent a 16-inch Dobsonian reflector for $35 for an entire night. You do not have to be a guest at the bed and breakfast to use the telescopes.

The Astronomer's Lodge at McDonald Observatory

This is where professional astronomers stay when doing research at McDonald Observatory. And, if one of the sixteen rooms is available, you can stay here, too, a rare opportunity to experience the on-the-job lifestyle of a professional astronomer. Because astronomers work during the night and sleep during the day, you have to follow the rules which basically say: keep it dark and quiet. The rooms are appropriately decorated in an astronomical motif with yellow stars and Moons. There are no TVs in the rooms, an omission intended to encourage guests to go watch the stars. The rooms have private baths, but everything else is communal, including the entertainment room and the dining room where guests can mingle with the pros. The rooms are $80 for a single and $128 for a double. Rates include two freshly prepared meals a day, and breakfast food is available around the clock. A separate director's house offering a fireplace, a kitchen, and a spectacular view is available for $250. (President George W. Bush stayed here while he was governor of Texas.) There is only one catch: to stay at the lodge, you must be participating in a special observing program on one of

> Website: www.as.utexas.edu/travel/mcdonald.php
> (McDonald Observatory)

the large telescopes or join up with the "Friends of McDonald" for $50. For further details and room reservations, consult the website.

If you want a more permanent astronomy-friendly living situation, then you can live in a community designed specifically for star-gazers. Light pollution is anathema to amateur astronomers, and these communities provide a dark sky sanctuary far away from city lights. The main rule that you have to live by is: Turn out the lights! It takes your pupils about half an hour to fully adjust to the dark conditions that astronomers crave. If exposed to white light, your pupils contract again. Thus, these communities have strict rules about using white light: no car headlights after dark, windows must be lined with light-blocking material, and outdoor lighting and flashlights must use dim red light—a color that doesn't affect the eyes the way white light does. There are currently three astronomy communities in the United States, but only one, Deerlick Astronomy Village, has property available as of early 2008. Evidently, the demand for this type of community far exceeds the supply, so enterprising astronomical entrepreneurs are sure to open up more communities in the future.

Arizona Sky Village

This residential community has more than 80 single-family lots with a minimum size of four acres. Residents may construct a single home and observatory on each lot with a minimum combined living area of 1,000 square feet. Amenities include a 4,000-square-foot Community Center with kitchen facilities and a Community Observatory featuring a remote CCD camera and a 30-inch computerized telescope. Telescopes are also available to rent. Future planned amenities include access to a robotic telescope and lectures by professional astronomers. Unfortunately, all the lots have been snatched up. However, reservations are being accepted for eleven "Interval Ownership" (what used to be called "time-share") adobe-style haciendas. Each 1,700-square-foot hacienda will be fully furnished with three bedrooms and two bathrooms. Each unit will be equipped with a 14-inch

> Website: www.arizonaskyvillage.com

computerized telescope in an observation courtyard. The preconstruction price is $10,000 per week. Arizona Star Village is located in the southeastern corner of Arizona off of Interstate 10 near the New Mexico border.

Chiefland Astronomy Village

This was the first astronomy community. Amateurs have been observing from this spot since 1985, and today more than 20 like-minded, star-gazing property owners have adjoining parcels of land. This enables them to cooperatively control light pollution. There are currently nine private observatories here with others on the drawing board. Although there isn't any land for sale here anymore, for $35 you can become a member of the Chiefland Astronomy Village Club and have access to the dark skies. The club maintains a large observing field with a shaded pavilion, flush toilets, and showers. There

> Website: www.chiefland.org

are nine RV outlets and forty regular electrical outlets around the perimeter of the observing field. Chiefland Astronomy Village is located near Chiefland, Florida, which is west of Gainesville.

Deerlick Astronomy Village

Located near the small town of Sharon in eastern Georgia, this is the newest of the astronomy communities. The first group of seventeen two-acre plots sold out in less than two years, but more are available. Small fifty-foot square plots that can be used for private observatories are available for leasing. The

village also includes a ten-acre hilltop observing field where stargazers

Website: www.deerlickgroup.com

are welcome to pitch a tent and set up their telescopes. Nonproperty owners can join the village for $35 and take advantage of the dark skies.

8. Attend a Star Party!

What do you call an event where a group of amateur astronomers, in a spirit
of astronomical fellowship and camaraderie, gather at a remote location far
away from hated city lights to watch the stars together? A "star party," of
course! Legend has it that the idea of a star party can be traced back to King
George III of England who exhibited a passionate fascination with astronomy. On nights when the real planets and stars were rendered invisible by
clouds, servants hung paper lanterns decorated with drawings from the trees
outside the royal palace so the king and his guests would have something to
look at through their telescopes. The first major star party in the United
States was the Stellafane Convention held near the town of Springfield, Vermont, in 1923. Today, hundreds of astronomy-related gatherings happen
every year across the United States. The gatherings usually last a few days
and nights, though some of the biggest affairs are annual events that stretch
throughout a week and attract hundreds or even thousands of enthusiasts.
Some attendees camp out on the ground or in tents, while others retire to
their RVs or commute from local motels.

So what do you do at a star party? Well, people bring their telescopes,
set them up in a field, and at night take turns looking through them.
Astrophotography and imaging are also popular activities. The larger parties
may include lectures, tours of nearby observatories or other astronomy-
related sites, swap meets, contests and raffles, and displays by commercial
vendors. Some participants exhibit their homemade telescopes, and prizes
are awarded for the best.

A few of the leading star parties in the United States are listed below.
Their websites are full of photos from previous events and provide details on
the next big party. For a more complete listing, go to the "Sky and Telescope" website and click on the "Event Calendar."

Stellafane, near Springfield, Vermont in July or August
    (www.stellafane.com)
Enchanted Skies Star Party, near Socorro, New Mexico in September or
    October (www.socorro-nm.com/starparty/index.html)

Winter Star Party, at West Summerland Key in the Florida Keys in
February (www.scas.org/wsp.html)
Texas Star Party, near Fort Davis, Texas in April or May
(www.texasstarparty.org/)
Riverside Amateur Telescope Makers Astronomy Expo, Riverside,
California over Memorial Day weekend
(www.rtmcastronomyexpo.org/)
Nebraska Star Party, near Valentine, Nebraska, in July or August
(www.nebraskastarparty.org/)

7. Train Like an Astronaut!

So you think you have the "right stuff" to be an astronaut? NASA gives you
the chance to test your mettle with the Astronaut Training Experience at the
Kennedy Space Center. This full-day program gives you a taste of the kinds
of training activities that real astronauts have to endure. The program begins
with an orientation and briefing by a former member of the astronaut corps
followed by several exercises that prepare astronauts for the physical challenges of space flight. At the Micro Gravity Wall, you are fitted with a har-
ness and weights that counter your own body weight so that you can climb
the wall with minimal effort. The 1/6 gravity chair uses a system of springs
and pulleys to simulate what it feels like to walk on the Moon. Finally, the
Multi-Axis Trainer spins and twirls you in all directions, thereby giving you
the sensation of hurtling through space. Be advised that the simulators have
height and weight restrictions. After lunch, you get a behind-the-scenes tour
of the Kennedy Space Center, including visits to the press site, the International Space Station Center, and the Space Shuttle launch pads. The day
culminates in a simulated Shuttle mission where participants work as a team
to launch the Shuttle into orbit, rendezvous with the International Space
Station, and deliver needed supplies and perform critical repairs. Depending
on the task assigned to you, you either use a realistically outfitted mission
control room or a full-scale mock-up of the Space Shuttle.

Participants in the Astronaut Training Experience must be at least 14
years old, and participants under the age of 18 must be accompanied by a
paying adult. The price for
the program is $250 per per-

Website: www.kennedyspacecenter.com

son. NASA also offers a two-day Family Astronaut Training Experience and a
customized Corporate Astronaut Training Experience. Advance registration
is required.

6. Watch a Space Shuttle Launch

If you want to see a Space Shuttle launch you had better hurry. The last Space Shuttle mission is scheduled for 2010. After that, there will be no manned launches for several years until the new Ares rockets start blasting off in 2014. The launch schedule is available online, but keep in mind that launches are often delayed because of weather or technical difficulties so be flexible. You can purchase a launch ticket online at the KSC website. The

Tomasz Szymanski/Shutterstock

Launch of the Space Shuttle from the Kennedy Space Center.

launch tickets run about $50 for adults and include admission to the visitor complex. You'll park at the visitor complex and board a bus to a viewing area about six miles from the launch site. The launch tickets are non-refundable, but they are issued for a particular mission, not a particular date, so if the launch is delayed, the tickets are still good as long as you haven't boarded the launch transportation bus. If you decide to buy a launch ticket, be sure to read and follow the procedures outlined in the online "Launch Information and Frequently Asked Question Packet."

Of course, public places near the KSC afford you an excellent view of the launch without paying NASA for the privilege. One of the best spots is Space View Park in Titusville, right across the river from the KSC and about ten miles from the Shuttle launch site. The park is on Broad Street at the Indian River. Get there at least an hour before launch and be prepared for a massive traffic jam after the launch. You can watch launches of small, unmanned rockets at the Wallops Flight Facility in Virginia.

5. Ooh and Aah at the Aurora!

Nature's most beautiful light show is the aurora which usually appears as a diffuse glow or as a shimmering curtain of multicolored light seen near the north and south polar regions. In the northern hemisphere, the aurora is often known as the northern lights because it appears in the northern part of the night sky. The more technical name is the *aurora borealis* originating from Aurora, the Roman goddess of dawn, and *boreas*, the Greek word for north wind. In the southern hemisphere, the southern lights appear in the southern part of the night sky and are called the *aurora australis*. The Latin word *australis* means "of the south."

The aurora is caused by charged particles, mostly electrons and protons, that are emitted continuously from the Sun. This stream of particles, called the solar wind, reaches the Earth after a two- to three-day journey and encounters the Earth's magnetic field. It is a basic fact of physics that whenever a charged particle moves through a magnetic field at an angle, a magnetic force is exerted on it. In this way, the Earth's magnetic field captures these particles and steers them along a spiraling path toward one of the Earth's two magnetic poles located within a few hundred miles of the north and south geographic poles. When the particles hit the atoms and molecules of gas in the Earth's upper atmosphere, the gases emit light. The combined light from billions of such collisions is what we see as the aurora. The colors

Roman Krochuk/Shutterstock

A view of the Northern Lights.

that appear in the aurora are determined by the kind of gases that are hit by particles. The two most common gases in the Earth's atmosphere are nitrogen and oxygen and so the colors produced by these gases are predominant. The red and green light is emitted by oxygen, and the blue and purple light by nitrogen. Photographing an aurora can be tricky because of the dim light and involves more than just pointing and clicking. Tripods and longer time exposures are needed. Filming an aurora requires specialized video equipment and is best left to the experts.

The aurora can be most frequently and easily seen within a 1,500-mile radius of the magnetic poles. Occasionally, during periods when the Sun is very active, the aurora can be seen from more temperate latitudes. One of the most dazzling auroral displays in recorded history happened in 1859 as a result of the Sun ejecting a large amount of matter from its corona. An account in the *New York Times* claimed that you could read at night by the light of the aurora.

The most convenient time to see the aurora borealis is during the months of September and October and from March through April. The aurora cannot be seen in the summer because the night sky is too bright, and during the winter the weather makes it inconvenient. Another factor to consider is the phase of the Moon. A full Moon overwhelms the light from the aurora, so the best viewing is near the new Moon phase. A number of commercial tour operators schedule "Northern Lights Tours" during these optimum periods. These tours are advertised in popular science magazines such as *Sky and Telescope* and *Astronomy*. You can also visit the websites of these magazines and peruse their listing of upcoming astronomy tours. (By the way, there are other types of astronomy-related tours including trips to visit archeoastronomical sites, use foreign observatories, or see an eclipse.)

Websites: www.skyandtelescope.com;
look under "astronomy travel"
www.astronomy.com;
look under "trips and tours"

4. Witness a Total Solar Eclipse

Commercial travel companies are offering an increasing variety of special interest tours, and astronomy is one of the more common of these interests. The most popular type of astronomy-related tour centers on the most spectacular event in the sky: a total eclipse of the Sun. A total solar eclipse happens when the disk of the Moon passes directly in front of the disk of the Sun and blocks out the sunlight. One might expect that a solar eclipse would happen about once every month when the Moon is in the "new moon" position directly between the Earth and the Sun. But the orbit of the Moon is tilted at an angle of about five degrees with respect to the plane of the Earth's orbit so the Moon usually passes a little bit above or below the Sun. Even if the alignment is right, a total eclipse does not always occur because the distance between the Earth and the Moon varies, which, in turn, changes the apparent size of the Moon as seen from Earth. If the disk of the Moon is slightly smaller than the disk of the Sun, then the Sun's light is not entirely blocked out; instead, a ring of sunlight is visible. The mathematical name for a "ring" is an *annulus* so this is known as an annular eclipse. But

A view of a total solar eclipse.

if the distances and angles are just right, then the Moon is exactly the same size as the Sun and moves directly in front of the Sun; a total solar eclipse occurs. A total eclipse of the Sun can be seen from somewhere on the Earth every few years. Although a total eclipse lasts only a few minutes, it is one of nature's most awe-inspiring sights. When the last little bit of sunlight disappears behind the Moon, the stars come out in the daytime, and the gossamer halo around the Sun known as the solar corona is visible. The fact that, from the vantage point of Earth, the Moon and the Sun appear to be the same size is purely coincidental. If intelligent beings on other planets are circling other stars, then it is highly unlikely that they have witnessed a total eclipse of their Sun. A total eclipse of a star is very rare event throughout the entire galaxy.

The path of a total solar eclipse cuts a long but narrow path across the surface of the Earth. Typically, the path is about 10,000 miles long, but only about 100 miles wide. If you want to see it, you will probably have to travel to a foreign destination. The next total eclipse visible from somewhere in the continental United States is in 2017. Of course, you don't have to sign up with a tour to see an eclipse. You are free to make your own arrangements, but there are advantages to taking a tour. First, most eclipse tours employ at least one professional astronomer who gives lectures about various eclipse-related topics and can offer advice on the best way to photograph the eclipse. A meteorologist also accompanies the tour to help make last-minute decisions on the best position for observing. Second, the eclipse tour operators book most available hotel rooms years in advance so it may be difficult for individuals to find accommodations, especially in remote locations where rooms are scarce. Finally, a group tour package to a foreign destination is usually cheaper than making your own arrangements.

The absolute best way to see an eclipse is not on a land-based tour, but on a cruise. Why? Although it is more expensive, a cruise ship can quickly move in any direction to position itself in a cloud-free location to view an eclipse, whereas on a land-based tour, you are restricted to a road, which may not be going in the direction you need to go to find clear skies. Experienced eclipse chasers say that if you are on the ground, you have a fifty-fifty chance of actually seeing the eclipse, but if you're at sea, you will nearly always be able to find clear skies. These tours are advertised in the astronomy magazines and websites listed in the previous entry.

3. Blast Your Ashes into Space!

Who can forget the moving scene in *Star Trek: The Wrath of Khan* where the body of Mr. Spock is launched into space to the tune of "Amazing Grace" played on the bagpipes? I admit it—I cried like a baby. Now you, too, can have your ashes blasted into space for a price that is affordable for most Americans. To keep the cost down, only a few grams of your cremated remains are launched into space in a capsule about the size of a tube of lipstick. Launching an intact human corpse into space would be prohibitively expensive.

The first space burial in history took place on April 21, 1997, when a modified Pegasus rocket was launched into Earth orbit from an aircraft and orbited the Earth for over five years. The rocket carried the ashes of 24 people, including the 1960s icon and drug proponent Timothy Leary and *Star Trek* creator Gene Roddenberry. In 1998, the second space burial carried the remains of planetary scientist Dr. Eugene Shoemaker on board the NASA Lunar Prospector probe, which orbited the Moon for eighteen months and then crashed into the lunar surface. Shoemaker is best known for his codiscovery of Comet Shoemaker-Levy 9, which slammed into Jupiter in 1994. He was also slated to be the first scientist to walk on the Moon but was disqualified after being diagnosed with a medical condition. In 2006, the remains of Clyde Tombaugh, discoverer of Pluto, were launched aboard NASA's New Horizons spacecraft, which will journey beyond Pluto's orbit. In 2007, the remains of both Gordon Cooper, one of the original Mercury 7 astronauts, and James Doohan, *Star Trek*'s "Scotty," were launched on a brief suborbital flight and recovered.

Currently, a company called "Space Services, Inc." is the only company that offers space burials, although as rocket technology expands into the private sector, more companies are expected to enter the market. Space Services, Inc., acquired the assets of a company called Celestis, Inc., which launched four burial flights between 1997 and 2001. The "Memorial Space-flights" are made possible through contracts that Space Services has with NASA and commercial space launch providers.

Space Services offers several memorial packages, the least expensive of which, the "Earth Rise Service" package, launches one gram of cremated remains into space and returns them safely to Earth where they are re-covered and returned to the family as a keepsake. All this for the low, low price of $495! Loved ones may attend the launch, and the price includes a professionally produced DVD of the memorial service and launch and a personalized online memorial. The "Gemini" option launches the mixed ashes of two people; the ashes of the first participant are kept in a safe deposit box until the second participant's "time of need." In case the mission is not successful, you are guaranteed a spot on the next

Website: www.spaceservicesinc.com

scheduled launch. Other, more expensive options include a launch into Earth orbit starting at $1,295, a launch into lunar orbit starting at $12,500, and a launch into deep space starting at $12,500.

2. Experience Weightlessness!

The Zero Gravity Corporation is the first commercial company to offer flights during which passengers can experience weightlessness. The company's business plan identifies five different markets for their service: adventure travel, corporate and incentive flights, the film and entertainment industry, research and education, and government agencies. Since its first flight in 2004, the company has flown more than 2,500 passengers aboard one hundred flights. The flights leave from the Kennedy Space Center in Florida and McCarran International Airport in Las Vegas. The planes, dubbed "G-FORCE-ONE," are specially modified Boeing 727s and can hold up to 35 passengers along with 6 crew members. The $3,500 ticket price includes a preflight briefing, flight suits, the 90- to 100-minute flight, and a "re-gravitation" party upon your return. In addition, each trip has a staff photographer onboard ready to snap photographs and the aircraft is equipped with High Definition video cameras to record your adventure on a DVD. A number of celebrity passengers have enjoyed the flight, including Martha Stewart, Penn Jillette, and *Apollo 11* astronaut Buzz Aldrin. Theoretical physicist Stephen Hawking, one of the world's leading experts on gravity who has been confined to a wheelchair for much of his life, fulfilled a life-long dream when he took the flight in 2007.

During each flight, the aircraft executes 15 parabolic arches between an altitude of 24,000 and 32,000 feet. The downward portion of each arch results in about 30 seconds of reduced gravity or weightlessness. When the plane is ready to execute its parabolas, the passengers lie flat on their backs on the floor of the 90-foot padded floating area. As the plane climbs into position for its dive, the passengers at first feel pressed into the floor with a force of 1.8 times that of normal gravity. The heavy feeling lasts for only a few seconds until the plane reaches the crest of the arc and begins its dive. Suddenly, the passengers begin to float up gently into the air. To ease your way to weightlessness, the first couple of parabolas simulate Martian gravity, about one-third of Earth's gravity, the next two or three simulate lunar gravity, about one-sixth of Earth's gravity, and the rest are done at zero gravity.

NASA has used this method to acclimate astronauts to the zero gravity environment of space for decades. The NASA astronaut training aircraft is officially known as a KC-135 Reduced Gravity Aircraft, but is better known by its unofficial name, the "Vomit Comet." The scenes requiring weightless conditions in the movie *Apollo 13* were filmed inside the Vomit Comet. The descriptive nickname for the NASA aircraft brings up (no pun intended) a natural question: Do passengers onboard the Zero-G flights get motion sickness? The company reports that only a small fraction of its passengers get sick. Why? Research indicates that 25 parabolas is the point at which many people start to feel sick, so the company limits the number of parabolas to 15. On a typical mission, NASA's Vomit Comet may do 40 to 80 parabolas.

So how is it that taking a steep dive in an airplane allows one to experience weightlessness? To understand this, suppose you took a set of bathroom scales to the cabin of an elevator, set the scales on the floor, stepped on the scales, and read your weight. The scales read your weight because it is being squeezed between your feet and the floor. Now press the button to go to the top floor. The elevator accelerates upward and you have the sensation of feeling heavier than normal—of being pressed down onto the floor of the elevator. If you watch the scale, it reads a higher than normal weight. Does this mean you have literally gained weight by pushing the "up" button? Of course not. The scale shows a higher weight because the floor is now accelerating upward against the scales and your feet. Thus, the scale is being squeezed more and so your "apparent weight" is greater. When the elevator has reached its cruising speed, your weight returns to normal. The opposite effect happens when the elevator slows down or decelerates as it approaches its destination. Now the floor of the elevator is not pushing up as hard and the bathroom scales read less than your normal weight. Now let's consider the extreme case where the cable breaks, all the safety systems fail, and the elevator is in "free fall" toward the Earth. If you were unfortunate enough to find yourself in this sad situation, what would the bathroom scale read? Zero, because now the scale isn't being squeezed at all. And, if you were to give a little push with your toes off the scales, you would find yourself floating around inside the cabin of the elevator. You are now in a state of "apparent weightlessness." You may have experienced this although in a safe, controlled way, if you have ever ridden in an amusement park ride that simply takes you up and drops you, a la the "Tower of Terror" ride at

Websites: www.gozerog.com
www.sharperimage.com
Telephone: 800-ZEROG800 ext. 0

DisneyWorld. For just a couple of terrifying seconds, you experience weightlessness as you hurtle toward the Earth. In Zero-G flights, you are falling in an airplane, but the physics is basically the same.

The Zero Gravity Corporation has established a partnership with the Sharper Image Corporation to market and sell seats on its flights. See the Sharper Image website for details.

1. Travel into Space—for a Price!

Journeys into space for the average person are no longer merely fodder for science fiction stories set in the far distant future; the future is now, but you had better start saving your pennies. On April 28, 2001, Dennis Tito, an American aerospace engineer and investment management entrepreneur, blasted off with his Cosmonaut compadres aboard a Russian Soyuz rocket and rendezvoused with the International Space Station where he spent a week performing a few experiments and cavorting around in space. Because he had to pay for his ride out of his own pocket, Tito is considered to be the first space tourist in history. The trip reportedly cost him a cool $20 million. But even at this price, the Russian Space Agency is fully booked until 2009. As of early 2008, four other space tourists have paid for the privilege of flying into space: South African computer millionaire Mark Shuttleworth in 2002; American scientist and entrepreneur Gregory Olsen in 2005; Iranian American business woman Anousheh Ansari in 2006; and Hungarian American Microsoft computer billionaire Charles Simonyi in 2007.

Several of these space voyagers have expressed their disapproval of the label "space tourist," preferring instead the term "private space explorer" or "independent researcher." The NASA Public Affairs Office has used the term "Spaceflight Participant." All five space tourists have flown to the International Space Station aboard the Russian Soyuz spacecraft. Why have all the flights been on Russian spacecraft? Unlike NASA, the Russian space program, starved for cash, uses the money collected from the space tourists to fund the program. NASA, along with private space travel organizations, have done market studies that show a potentially sizeable market for space tourism, a market that is sure to expand as the cost of putting a person into space contracts. Adventure tourism with destinations to remote locations such as Mount Everest and Antarctica has proven to be a profitable market, even with packages priced as high as $100,000.

In October 2004, Burt Rutan, an American aerospace engineer and innovator, won the Ansari X Prize, a competition awarding a cash prize of $10 million to the first private organization to launch a reusable manned spacecraft into space twice within two weeks. Billionaire Richard Branson and his Virgin Group formed a company called, appropriately enough, Virgin Galactic, and has contracted with Burt Rutan to design and build a six-passenger vehicle that will safely and routinely fly to an altitude of 68 miles, slightly beyond the internationally defined boundary between Earth and space. Within a month after the announcement of the contract, Virgin Galactic had received more than 7,000 inquiries from people interested in acquiring tickets. Passengers will be able to see below them the curved surface of planet Earth and above them the twinkle-free stars against the blackness of space. For kicks, the passengers will experience a few minutes of weightlessness. The flights will last two-and-a-half hours and reach a top speed of Mach 3. The ship, named *SpaceShipTwo* or the *VSS Enterprise* (*SpaceShipOne* was the original design that won the X-Prize), will be launched from an aircraft flying at 50,000 feet, rather than from the ground like the Space

Shuttle. By using a gentle "feathering" technique for reentry, *SpaceShipTwo* will avoid any extreme heating during descent, and thus the ship will not require a heat shield like the Shuttle. Passengers will receive three days of training before launch, and the trip will initially cost $200,000. The inaugural launch is set for 2008 with regular flights beginning in 2009.

Virgin Galactic is not alone in the infant space tourism industry. Perhaps the leading company in the business is Space Adventures, Inc. This full-service space tourism agency can arrange everything from Space Shuttle launch tours to Soyuz flights to the International Space Station. This agency made travel arrangements for Dennis Tito and the other early space adventurers mentioned above. Space Adventures has on its advisory board several former NASA astronauts, including Buzz Aldrin. In 2005, Space Adventures announced that they were working with the Russian Space Agency to fly two passengers on a mission around the Moon for $100 million per seat. Constellation Services International (CSI), a rival company, is working on offering a two-week space excursion: a week-long stay at the International Space Station followed by a week-long trip around the Moon.

Robert Bigelow, an American motel tycoon, has formed Bigelow Aerospace, a startup company that is working on inflatable space habitat modules that could be joined together to form a commercial space station. The company has already launched test modules and plans to launch the space station project, called

Websites: www.spaceadventures.com
www.virgingalactic.com

Nautilus, in 2010. Bigelow Aerospace has offered a $50-million prize to the first company that can build a reusable spacecraft capable of transporting passengers to the Nautilus space station.

A STATE-BY-STATE LIST OF SITES

Alabama

Marshall Space Flight Center, Huntsville

Arizona

Kitt Peak National Observatory, Tucson
Lowell Observatory, Flagstaff
Meteor Crater
Museum at the Center for Meteorite Studies, Tempe

California

Ames Research Center, Moffett Field
Dryden Flight Research Center/Edwards Air Force Base
Griffith Observatory, Los Angeles
Hubble House, San Marino
Jet Propulsion Laboratory, Pasadena
Lick Observatory, San Jose
Mount Palomar Observatory, San Diego County
Mount Wilson Observatory, Pasadena

Colorado

Chimney Rock Archaeological Area
Hovenweep National Monument, Cortez
Mesa Verde National Park

Florida

John F. Kennedy Space Center, Cape Canaveral

Hawaii

Keck Observatory

Illinois

Adler Planetarium, Chicago
Cahokia Mounds

Indiana

Armstrong Statue, West Lafayette
Gus Grissom Memorial, Mitchell

Kansas

Kansas Cosmosphere and Space Center, Hutchison

Maryland

Goddard Space Flight Center, Greenbelt

Massachusetts

Maria Mitchell House, Nantucket Island

Mississippi

Stennis Space Center

Nevada

Area 51

New Mexico

Chaco Canyon National Historical Park
Meteorite Museum, Albuquerque
National Solar Observatory
New Mexico Museum of Space History, Alamogordo
Roswell
Very Large Array, Socorro

New York

Hall of Meteorites, American Museum of Natural History, New York City
Rose Center for Earth and Space Science, New York City

New Jersey

Einstein's House, Princeton

North Dakota

Medicine Wheel Park, Valley City

Ohio

Armstrong Air and Space Museum, Wapakoneta
Glenn Research Center, Cleveland
John and Annie Glenn Historic Site, New Concord

Pennsylvania

David Rittenhouse Birthplace, Philadelphia

Puerto Rico

Arecibo Radio Observatory

Texas

Johnson Space Center, Houston
McDonald Observatory, Fort Davis
Odessa Meteor Crater
Oscar E. Monnig Meteorite Gallery, Fort Worth

Virginia

Udvar-Hazy Center
Virginia Air and Space Center, Hampton
Wallops Flight Facility, Wallops Island

Washington, D.C.

Einstein Memorial
National Air and Space Museum
National Museum of Natural History, Meteorite Gallery
U.S. Naval Observatory

West Virginia

National Radio Astronomy Observatory

Wisconsin

Deke Slayton Memorial Space and Bike Museum, Sparta
Yerkes Observatory, Williams Bay

Wyoming

Big Horn Medicine Wheel

INDEX

ABOUT THE AUTHOR

DR. DUANE S. NICKELL teaches physics at Franklin Central High School in Indianapolis, Indiana, and is an associate faculty member at Indiana University/Purdue University at Indianapolis. He holds a doctorate in education from Indiana University and has won numerous teaching awards, including the prestigious Presidential Award for Excellence in Science and Mathematics Teaching, the nation's highest honor for science and mathematics teachers.